P9-DYX-730

CITRUS

A COOKBOOK

BY FORD ROGERS · PHOTOGRAPHS BY ERIC JACOBSON

DESIGN BY LESLEY EHLERS

A FIRESIDE BOOK
PUBLISHED BY SIMON & SCHUSTER INC.
NEW YORK LONDON TORONTO SYDNEY TOKYO SINGAPORE

A RUNNING HEADS BOOK

FIRESIDE
Simon & Schuster Building
Rockefeller Center
1230 Avenue of the Americas
New York, New York 10020

FIRESIDE and colophon are registered trademarks of Simon & Schuster Inc.

CITRUS: A COOKBOOK
was conceived and produced by
Running Heads Incorporated
55 West 21 Street
New York, New York 10010

Editor: Rose K. Phillips
Managing Editor: Jill Hamilton
Production Manager: Linda Winters
Production Associate: Belinda Hellinger

1 3 5 7 9 10 8 6 4 2

Library of Congress Cataloging in Publication Data

Rogers, Ford Barker III.
Citrus, a cookbook / by Ford Barker Rogers III : photographs by
Eric Jacobson.
p. cm.
"A Fireside book."
"A Running Heads book"—T.p. verso.
ISBN 0-671-74534-4
1. Cookery (Citrus fruits) I. Title
TX813.C5R64 1992
641.6'43—dc20 91-34332
CIP

Typeset by Trufont Typographers, Inc.
Color separations by Hong Kong Scanner Craft Company, Ltd.
Printed and bound by Tien Wah Press (Pte.) Ltd.

For my mother and father, my grandmother Lois Everson, and Ruth Boston, who taught me to love food

AUTHOR'S ACKNOWLEDGMENTS

Special thanks to Steve Smith for his inspiration, support and love, and for transferring my scribbles to disk. For their love and encouragement, thanks to my family and friends, particularly Martie LaBare and Michael McEvoy, Fred and Beth Rust, and John Morse for tasting recipes, and Alan Mittelsdorf for his friendship and support over the years.

Thanks to Carla Glasser and John Morse (again) at the Nolan-Lehrer Group for getting me started. Special thanks to Sydny Miner, Clare Wellnitz, and Mary Kapp at Simon & Schuster, and Marta Hallett, Ellen Milionis, Jill Hamilton, and Rose K. Phillips at Running Heads, and Linda Greer, as well. Finally, thanks to Dexter Samuel for his good-humored help and to David Blackburn, Jeff Romano, and Tina Mae Jones of Native Farms for sensational citrus.

PHOTOGRAPHER'S AND DESIGNER'S ACKNOWLEDGMENTS

We would like to thank the following people for their support during this project: the Jacobson and Ehlers families, Judy Devine for her inspiration and love, Dexter Samuel, Jill Bock, Beth Farb, Michele Cohen, Kim Kelling, Tina Klem, Jon Roemer, Robin Van Loben Sels, and Linda Winters.

Thanks also to the many people and companies who loaned us their beautiful tablewares: Williams-Sonoma, Pottery Barn, Laura Beck at Platypus, Kaija at Ceramica, Jim Mellgren at Dean and Deluca, Pauline Kelley and Lorena Sita at Zona, John P. Gilvey and Michael Benzer at Studio Art Glass, Kathy Dahlberg at Bayer Glass Studios, and Elizabeth MacDonald for her fabulous tilework.

CONTENTS

INTRODUCTION
8

ONE
BREAKFAST AND BRUNCH

Pink Grapefruit Poached in Sauternes
14
Stewed Kumquats and Strawberries
16
Orange-Banana Muffins
18
Lemon and Allspice Muffins
18
Tangerine-Pecan Scones
20
Navel Orange Waffles with Blueberry Sauce
22
Orange French Toast
24
Buckwheat Crepes with Clementines
26

TWO
APPETIZERS, SOUPS AND SALADS

Grilled Pink Grapefruit and Pork Skewers
30
Lime Soup
32
Lemon-Parsley Soup
34
Curried Orange Soup
36
Salad of Beets, Mache and Clementines
38
Ugli Fruit Caribbean Fish Salad
40
Chicken Salad with Tangerines and Red Onions
42
Pickled Citrus Shrimp
44

THREE
MAIN COURSES

Lemon Fettucine with Peppered Shrimp
48
Fried Scallops with Lime and Garlic
50
Grilled Shark Steaks with Citrus Salsa
52
Baked Fish with Ugli Fruit
54
White Grapefruit-Marinated Cornish Hens
56
Chicken with Lemon and Olives
58
Soba with Fennel, Chèvre and Kumquats
60
Limed Spareribs
62
Pork Medallions with Mandarin Oranges and Cranberries
64
Veal Chops with Blood Oranges
66
Beef with Tangerines
68

FOUR
RELISHES AND CONDIMENTS

Pickled Lemons
72
Citrus Marmalade
74
Kumquat and Ginger Preserves
76
Lime and Tomato Relish
78
Tangelo and Sweet Onion Relish
78
Tangerine-Cranberry Chutney
80
Orange Butter
80

FIVE
DESSERTS

Vermont Orange Ambrosia
84
Chocolate-Dipped Citrus Sections
86
Minted Ruby Grapefruit Ice
88
Tangerine Sorbet
88
Lemon Snaps
90
Chocolate-Orange Marmalade Brownies
92
Orange-Coconut Custard
94
Lime Mousse Pie
96
Frozen Blood Orange Soufflé
98
Lemon-Nutmeg Cheesecake
100
Lemon Steam Cake with Blood Orange Sauce
102
Mandarin Orange Upside-Down Cake
104

SIX
BEVERAGES

Pink Lemonade
108
Iced Orange Coffee
108
Limeade Fizz
110
Citrus Frappé
110
Hot Spiked Lemonade
112
Lemon-Thyme Tea
112
Sherried Pink Grapefruit Juice
114
White Grapefruit-Pineapple-Rum Punch
114
Blood Orange Liqueur
116

RECIPE LIST
118

RECIPE LIST BY CITRUS
119

SOURCES
120

INTRODUCTION

To many people, the uses for citrus fruits extend only to drinks and to a few familiar dishes. For centuries, however, cooks in cultures all over the world have found that the ways to use oranges, lemons, grapefruits, limes, and more unusual varieties—such as ugli fruit and kumquats—are practically limitless.

Growing up in Florida, both a citrus and seafood center, I learned, early on, the indispensability of citrus when serving all kinds of seafood—and that serving iced tea without a slice of lemon is tantamount to blasphemy. As a child, I travelled to Key West for my first real Key lime pie and was surprised to find it was *yellow* instead of green. I have vivid memories of climbing my family's backyard orange tree, which would fill the air with the sweet scent of its blossoms in springtime. My family would trade our oranges for the giant yellow grapefruits our next-door neighbors grew, and we always had plenty of lemons on hand for lemonade to make the hot, humid summers more bearable. Like many kids, my first commercial venture—the lemonade stand—owed everything to this versatile crop. Fresh citrus fruits formed the basis of the many fruit salads, refreshing main dishes, and even the home remedies (my mother favored grapefruit "punch" for laryngitis) we enjoyed year-round.

As I got older, it didn't surprise me to find that other cultures carried the love of citrus even further. I've enjoyed my encounters with citrus in cuisines from around the world. Caribbean cooks offer such wonderful dishes as seviche and a citrusy Christmas pudding, not to mention their citrus-infused signature planters punch and tropical cocktails. Grecian fare makes ample use of the lemon, in their avgolemono soup and such staples as squid with lemon and garlic. Asian cultures, too, rely on citrus to add diversity to their cuisines. Most everyone is familiar with beef with orange flavor, but there are also many other interesting citrus dishes to enjoy, such as Thai lime yum dressing (yum is a type of salad). Marvelous and inventive dishes such as these have inspired me to create the citrus recipes for this book.

Archeobotanists believe that a common ancestor of all of today's citrus fruits sprang up in India, or possibly in the Tigris and Euphrates Valley, 80 centuries ago. Crossbreeding refined and differentiated various forms of citrus, and these were further distributed through the known world. As this happened, the types of fruit and their uses became as varied as their locales.

Grown in Mesopotamia for their beauty and scent and in Egypt for their use in embalming, citrus fruits have been used, at various times, as aphrodisiacs, as cures for fever or colic, or as a means of protection against poisons.

Romans used a certain variety of citrus to keep moths from their woolens, while they relied on sour grapes to add tartness to their food (until Persian slaves introduced them to citrus's culinary uses). In the Dark Ages—a time when eating fresh fruit was considered by many to be harmful to the body—lemons were thought to be poisonous.

The discovery of scurvy prevention through citrus was made long before it was understood. In fact, the first citrus in the Western Hemisphere probably arrived with Christopher Columbus. The seeds quickly spread throughout the islands of the Caribbean, making the islands medical way stations for sailors. By the time Ponce de Leon arrived in Florida in 1513, sailors on Spanish ships were required to carry 100 citrus seeds for planting wherever they landed. By the 19th century, citrus had become a major industry in Florida.

Among modern-day inventive uses for citrus that I've encountered are employing lemon juice as "invisible ink" and the old fisherman's trick of using lemon juice to remove fish scents after cleaning the catch of the day. Citrus flowers, with their gorgeous scents, of course, are also potent ingredients in potpourri and other fragrant household blends.

But the most rewarding way to use citrus is, without question, as a culinary ingredient. You'll find recipes here that use navel and blood oranges; white, pink, and ruby-red grapefruits; tangerines or mandarin oranges; and lemons, limes, clementines, tangelos, and kumquats. Today, many of these fruits are available year-round, but others are still seasonal treats to be searched out and enjoyed but briefly. All are abundant in vitamin C.

Oranges in one variety or another are with us all the time. Some varieties, such as blood oranges, are usually around only in the winter and spring; others, like navels, are around most of the year. Valencias, a late variety, are available from the end of spring through the summer.

When selecting oranges—as with any citrus fruit—look for smooth, firm skins, free of soft spots. Pick a fruit that feels heavy for its size.

The exception to the smooth-skin rule is the unfortunately named "ugli fruit." I think of it as the Shar Pei of citrus fruits, with a loose and wrinkled skin that hides what is, perhaps, the best-kept secret on the citrus market. Uglis are sweet, delicious, and just about the juiciest fruit around. Originally a hybrid of a grapefruit and a tangerine, the ugli grows in the Caribbean and is available from December through May.

One of the ugli's ancestors, the tangerine, is also called a mandarin orange. Though the terms are interchangeable, smaller tangerines often end up labelled "mandarins." Tangerine skin doesn't have the bitter layer of white pith that surrounds other citrus fruits, but rather a few strands of it between the skin and the fruit. This makes the skin of tangerines particularly delightful and useful in cooking, such as with Tangerine-Pecan Scones. Throughout the book, zest is used to describe the colored part of the peel used in the recipe; however; with the tangerine, the entire peel is used. Of all citrus fruits, the tangerine's skin tastes most like the fruit it encloses, and it doesn't need the blanching often required with the others to remove the bitter taste. This delightful fruit is sweet and tasty, as Chinese cooks have known for centuries, and can be used fresh or dried. When selecting tangerines, look for those with deep color and loose skin for easier peeling.

The ugli fruit's other ancestor, the grapefruit, has had a hard time escaping its typecasting as a breakfast fruit. Available in traditional white and pink, as well as the more recent ruby-red variety, it is a far more versatile fruit than many people believe. Grapefruit makes thoroughly thirst-quenching drinks and palate-cleansing ices, as well as being a spectacular salad ingredient. If you insist on relegating grapefruit to the breakfast table, try a little salt on it, instead of sugar. My grandmother always said that salt cut the tartness much more efficiently than sugar. To my own taste, I find it sweet enough—particularly the ruby reds—with no additions. You can probably find grapefruit in your market all year, but they'll be freshest and in greatest abundance from January through April. Choose fruit that has a nice luster and is well-shaped and heavy.

Clementines are a cross between mandarins and bitter Seville oranges, but the taste is anything but bitter. They are sweet and easy to peel and section, and, best of all, seedless. They're the "little darlings" of the citrus trade. Unfortunately, clementines are available for only a relatively brief period in the winter. Smaller even than the clementine is the kumquat, which some consider the stepchild of citrus, with its reputation as a sour pucker-upper. These days, through cultivation, this stepchild seems to be something of a "Cinderella." It can be surprisingly sweet, particularly the skins. Some people even eat them for the purpose of squelching cravings for sweets. Traditionally limited to preserves or candied, the kumquat is coming into its own by lending its piquant flavor to fish and vegetable dishes. Buy kumquats that are firm and shiny, without blemishes or soft spots. If the green leaves are still attached, that's a good clue to the fruit's freshness.

Lemons and limes are available all year, but their prices vary with the season. Avoid shrivelled fruits or those with very soft or very hard skins. Lemons with green-tinged skins and the greenest limes are the tartest.

You'll find many varieties of lemon today, but the small, round, thin-skinned ones tend to be juiciest. Limes, which require hotter, moister climates to thrive, are used in place of lemons in many parts of the tropics. There are two basic types of lime you'll find in the market: the small, yellowish Indian variety, and the larger, very-green Caribbean one.

In general, I prefer to use organic fruit when possible, particularly when using the peel or zest. If you use nonorganic fruit, always wash it thoroughly in warm, soapy water. Complaints about organic fruits being less attractive just don't hold up anymore. Of course, some fruits do appear preternaturally bright and colorful. And, that may well be because they've been dyed.

Usually, Florida citrus isn't dyed, and a bit of greenish color is acceptable.

Here are a few more tips for getting the most out of your ever-versatile citrus selections:

- Most citrus fruits will keep at room temperature for 2 to 5 days. They'll keep in the refrigerator for 2 to 4 weeks.
- Use lemon juice in place of salt (this is especially good for those on a low-salt diet) or in place of vinegar for a cool, clean taste.
- Add the juice of a lemon, instead of salt, to the water used to boil pasta.
- To avoid discoloration, add citrus juice—lemon is usually used, but any fruit with citric acid will do—to lightly colored fruits, such as apples, bananas, pears, or avocados, when cutting them up for serving or freezing.
- To prevent darkening, add any kind of citrus juice to the cooking water of lightly colored vegetables, such as cauliflower, turnips, or potatoes.
- Substitute 1 teaspoon of lemon juice for ½ teaspoon of cream of tartar in meringue recipes.
- Tenderize tough meat by marinating it in any kind of citrus juice or by spooning juice over the meat which has been pierced with a fork.
- Freshen poultry, after washing, by rubbing it with a cut lemon or orange.
- Add 1 teaspoon of lemon juice per pound of fruit when canning, if you are unsure of the fruit's acid content.
- Never use iron or aluminum cookware in preparing citrus, as it will react with the metal, resulting in a metallic taste and darkening of the food.

Citrus fruits can be found wherever warm winds blow and the tropical sun shines. For the rest of the world, these fruits are tropical ambassadors with their bright colors and fresh, cool tastes. They have become integral ingredients in the cuisines of the world. Some dishes, such as lemon meringue pie, duck à l'orange, and lime sherbet are very well known and have appeared in hundreds of cookbooks. You may not find these and other old favorites in this book, but you might make a new friend in these sometimes surprising ways for using citrus fruit in cooking.

I think you'll find citrus's clean, cool tastes—sometimes tart and sometimes sweet—infinitely adaptable. Citrus fruits can enhance virtually any other flavor without overwhelming it. They can be used with the most delicate tastes, such as chicken or cucumber, where they support the flavor, or with pronounced tastes, such as lamb or onion, where they come into their own, delightfully altering the other familiar tastes.

My sources for these recipes come from all over the world—and from my own experience and imagination. In all cases, though, my inspiration comes from the simple-to-use, delicious-to-taste citrus fruits.

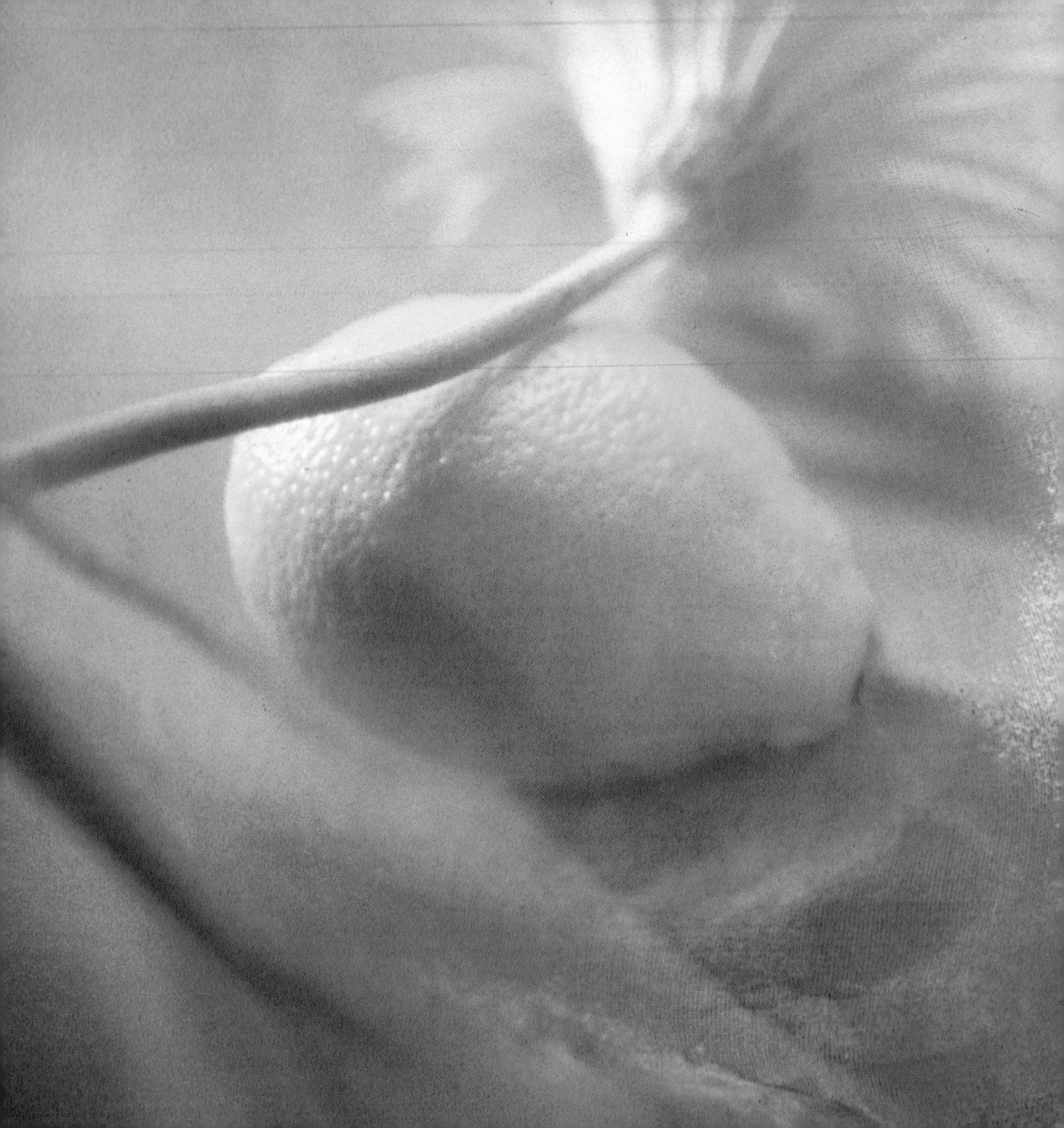

One · Breakfast and Brunch

Pink Grapefruit Poached in Sauternes

3 pink grapefruits
2 cups French sauternes
3 tablespoons brown sugar
mint leaves, for garnish

• Using a sharp paring knife, peel the grapefruits over a small bowl, cutting through the peel and pith, just into the flesh, then spiralling down and around from one end to the other, as if peeling an apple. Cut into the membranes, pop out the sections, and remove the seeds.
• In a 2-quart nonreactive saucepan, heat the sauternes and sugar almost to boiling. Add the grapefruit sections and simmer for 5 minutes. Remove from the heat and allow the grapefruit to cool a bit in the sauternes. Serve in the sauce while still slightly warm, or chill for serving later. Garnish with the mint leaves.

Serves 4.
Preparation time: 15 minutes.

Stewed Kumquats and Strawberries

- ¾ pound kumquats, washed and stemmed
- ¼ cup granulated sugar
- 5 cloves
- 1 pint strawberries, washed and hulled
- 2 tablespoons Grand Marnier

• Slice each kumquat crosswise, into 3 or 4 rounds. Remove seeds. Place in a 2-quart nonreactive saucepan with the sugar, cloves, and ½ cup water. Simmer for 5 minutes.

• Quarter the strawberries and add them to the pan. Simmer for 3 more minutes, then remove from heat. Let cool to room temperature.

• Add the Grand Marnier; cover and refrigerate several hours or overnight. Remove the cloves before serving. Serve as is or over vanilla ice cream.

Serves 4.
Preparation time: 30 minutes, plus several hours or overnight to chill.

Orange-Banana Muffins

¾ cup whole wheat flour
¾ cup rolled oats
½ cup unbleached, all-purpose flour
1 tablespoon baking powder
½ cup brown sugar
½ teaspoon salt
3 tablespoons wheat germ
1 teaspoon ground cinnamon
½ cup milk
½ cup freshly squeezed orange juice
1 large egg, lightly beaten
1 cup mashed bananas (about 2 bananas)
2 tablespoons grated orange zest
¼ cup unsalted butter, melted

• Preheat oven to 400°.
• Grease six 3-inch muffin pan cups.
• In a large bowl, combine the dry ingredients. In a small bowl, combine all the remaining liquid ingredients, mixing well. Add the liquid ingredients to the dry ones all at once and stir just until moistened. Spoon the mixture into the prepared cups, filling them ⅔ full.
• Bake approximately 25 minutes, or until they are golden brown and a toothpick inserted in the center of a muffin comes out clean.

Makes six 3-inch muffins.
Preparation time: 45 minutes.

Lemon and Allspice Muffins

2 cups unbleached, all-purpose flour
2 teaspoons baking powder
½ teaspoon salt
½ cup granulated sugar
1½ teaspoons ground allspice
1 tablespoon grated lemon zest
1 large egg, lightly beaten
⅔ cup milk
⅓ cup freshly squeezed lemon juice
¼ cup unsalted butter, melted

• Preheat oven to 400°.
• Grease twelve 2½-inch muffin pan cups.
• Sift the flour, baking powder, salt, sugar, and allspice into a mixing bowl. Add the zest and mix well.
• In a small bowl, combine the beaten egg with the milk, lemon juice, and melted butter.
• Add the liquid ingredients to the dry ones all at once, mixing only until well moistened (the batter should be a little lumpy). Fill muffin tins ⅔ full with the mixture. Bake 20 to 25 minutes, until golden brown.

Makes 1 dozen 2½-inch muffins.
Preparation time: 45 minutes.

Tangerine-Pecan Scones

2 cups unbleached, all-purpose flour
¼ cup granulated sugar
1 tablespoon baking powder
½ teaspoon baking soda
½ teaspoon salt
½ cup unsalted butter, frozen and cut into chunks
⅓ cup tangerine peel, with all white pith strands removed (2 to 3 tangerines)
¾ cup coarsely chopped pecans
¾ cup buttermilk
1 egg yolk

• Preheat oven to 400°.
• In the bowl of a food processor or by hand, combine the flour, sugar, baking powder, baking soda, salt, and butter. Process until the mixture resembles a coarse meal. Transfer to a bowl. Dice the tangerine peel into ¼-inch pieces. Add the peel and pecans to the mixture. Add the buttermilk and mix well, until the dough is stiff enough to handle.
• Turn out onto a well-floured board and knead gently until smooth, about 1 minute. Roll or pat the dough out to ¾-inch thickness. Cut with a 3-inch biscuit cutter or drinking glass dipped in flour. Arrange scones on greased baking sheet 1 inch apart. Re-form dough scraps into a ball, roll out and cut again until all dough is used.
• Brush tops with egg wash made of 1 egg yolk beaten with 1 teaspoon water. Bake for 15 to 20 minutes, until golden brown.

Makes 1 dozen 3-inch scones.
Preparation time: 40 minutes.

Navel Orange Waffles with Blueberry Sauce

1¾ cups unbleached, all-purpose flour
1 teaspoon baking powder
1 teaspoon baking soda
½ teaspoon salt
2 tablespoons granulated sugar
1¾ cups buttermilk
¼ cup unsalted butter, melted
3 tablespoons Cointreau or ½ teaspoon orange extract
2 eggs
2 tablespoons grated navel orange zest
vegetable oil
navel orange segments

• Preheat waffle iron.
• In a large bowl, combine the flour, baking powder, baking soda, salt, and sugar. Stir to mix. Add the buttermilk, butter, Cointreau, eggs, and zest and beat until smooth. When the waffle iron is ready, brush it lightly with oil and pour batter into the center until it spreads within 1 inch of the edge. Cover and bake following manufacturer's instructions.
• When done, carefully lift the cover and loosen the waffle with a fork.
• Serve immediately, garnished with orange segments and Blueberry Sauce (recipe follows), or hold briefly, uncovered, on a baking sheet in a preheated, warm oven, until all waffles are done.

Makes 4 to 6 waffles, depending on iron size.
Preparation time: 30 minutes.

Blueberry Sauce

1 pint blueberries, washed and de-stemmed
½ cup granulated sugar
1 tablespoon freshly squeezed lemon juice

• In a small saucepan over medium heat, combine the blueberries with sugar, lemon juice, and ¼ cup water. Bring to a boil and lower the heat. Simmer for 5 minutes. Serve warm.

Orange French Toast

¼ cup sour cream or plain yogurt
1 teaspoon ground cinnamon
2 tablespoons honey (preferably orange blossom honey)
2 large eggs
1¼ cup freshly squeezed orange juice
1 loaf French, challah, or Italian bread, 2 to 3 days old (the harder and crustier the better), sliced ½-inch thick
unsalted butter for frying

• In a small bowl, whisk the sour cream or yogurt with the cinnamon and honey. Add the eggs and beat until smooth. Gradually beat in the orange juice and continue to whisk until well combined and frothy.
• Dip the bread slices into the mixture a few at a time, soaking them thoroughly, but not so much that they fall apart.
• Melt about 3 to 4 tablespoons of butter in a skillet over medium-high heat and sauté the slices until golden brown on each side. Do not crowd the slices in the pan. Add more butter as needed.
• Serve immediately or hold briefly in a warm oven until all slices are done. Serve with Orange-Maple Syrup (recipe follows) or honey.

Serves 4.
Preparation time: 15 minutes.

Orange-Maple Syrup

1 cup maple syrup
¼ cup freshly squeezed orange juice

• In a small saucepan over low heat, combine the ingredients, stirring often until heated.

Buckwheat Crepes with Clementines

1 cup sifted buckwheat flour
½ cup sifted, unbleached, all-purpose flour
¼ teaspoon salt
1¾ cups milk
3 tablespoons unsalted butter, melted
3 eggs
4 clementines
⅓ cup granulated sugar
1½ cups mascarpone cheese at room temperature
vegetable oil

• Place the flours and salt in the bowl of a food processor or mixer. Process a few seconds to combine. With motor running, add the milk, butter, and eggs. Process just until smooth and blended. Set the batter aside for 30 to 45 minutes.
• Peel the clementines, removing all the white pith and reserving the peel of one of them. Separate and reserve the sections.
• Scrape the white pith off the reserved peel and finely chop it. Place it in a small saucepan with the sugar and ¼ cup water. Bring this mixture to a boil, stirring to dissolve the sugar. Add the clementine sections and lower the heat. Simmer for 5 minutes, then remove from the heat and set aside.
• Brush a crepe pan or 6-inch omelet pan with a paper towel dipped in oil. Heat over medium heat until a drop of water sputters on contact. Pour 3 to 4 tablespoons of batter into the pan and tilt the pan to coat the surface evenly with the batter.
• When the crepe is browned and can be shaken loose (after 3 to 4 minutes), turn it with your fingers or a spatula and brown the other side (2 to 3 minutes more). Slide the finished crepe onto paper towels to cool.
• If the batter is too thick, add a little milk to obtain the proper consistency.
• Stack the crepes as they cool. When ready to serve, spread a couple of tablespoons of the cheese across each crepe. Roll them up and place 3 crepes each on 4 serving plates. Spoon the clementines with their sauce over each plate.

Serves 4. Makes 12 crepes.
Preparation time: 1 hour 30 minutes.

Two · Appetizers, Soups, and Salads

Grilled Pink Grapefruit and Pork Skewers

1 whole pork tenderloin, about ¾ pound
¼ cup soy sauce
¼ cup firmly packed dark brown sugar
¼ cup freshly squeezed lemon juice
2 tablespoons dry sherry
1 tablespoon grated fresh gingerroot
1 garlic clove, minced
1 pink grapefruit

• Cut the pork into 1-inch cubes and place them in a bowl. Combine the remaining ingredients, except the grapefruit, and pour them over the pork. Cover and marinate for 1 hour.
• Slice the unpeeled grapefruit into 12 sections and halve each section.
• Thread the pork cubes alternately with the grapefruit pieces (piercing them through the skin) onto 8 skewers. Reserve the marinade.
• Grill over hot coals or broil, turning and basting often with the reserved marinade, for about 15 minutes, or until cooked thoroughly.

Serves 4.
Preparation time: 1 hour 30 minutes.

Lime Soup

2 teaspoons lime zest
½ cup freshly squeezed lime juice
1 medium jalapeño pepper, seeded and coarsely chopped
2 cucumbers, 7–8 inches long, peeled, seeded, and coarsely chopped
2 cups seedless white grapes
¾ cup coarsely chopped scallions
¼ cup fresh coriander leaves
½ cup coarsely chopped green pepper
¼ cup golden tequila
thin slices of lime for garnish

• Puree all the ingredients, except the last 2, in 2 batches, using half the lime juice in each batch, in a food processor or blender until smooth.
• Chill thoroughly. Add the tequila, mix well, and serve in chilled bowls or cups. Garnish with lime slices.

Serves 4.
Preparation time: 20 minutes.

Lemon-Parsley Soup

2 tablespoons unsalted butter
6 cups loosely packed Italian, flat-leafed parsley leaves (about 4 bunches)
1 teaspoon ground coriander
2 garlic cloves, crushed and peeled
3 cups rich chicken broth
¼ cup freshly squeezed lemon juice
1 tablespoon grated lemon zest
2 cups plain yogurt or sour cream
salt and freshly ground black pepper
parsley leaves for garnish (optional)
lemon zest for garnish (optional)

• Heat the butter in a 2-quart non-reactive saucepan over medium heat. Add the parsley leaves and cook, tossing frequently, until the leaves are completely wilted and coated with butter (about 3 to 5 minutes). Add the coriander and garlic and sauté for 2 more minutes. Add the broth, lemon juice, and lemon zest. Bring the mixture to a boil, then lower the heat and simmer for 10 minutes.
• Remove the soup from the heat and puree in batches. Wipe out the pan and return the soup to it. Add the yogurt or sour cream and mix thoroughly.
• If you are serving the soup cold, let it cool, then cover and refrigerate it until it is well chilled.
• If you are serving it hot, heat the soup through, but do not let it boil.
• When serving, season with salt and pepper to taste and garnish with parsley and lemon zest, if desired.

Serves 4.
Preparation time: 25 minutes.

Curried Orange Soup

1 tablespoon unsalted butter
1 medium onion, chopped
1 pound carrots, peeled and chopped
1 large garlic clove, peeled and crushed
3 cups chicken broth
½ teaspoon ground turmeric
3 whole cloves
1 teaspoon coriander seeds
12 black peppercorns
½ teaspoon cumin seeds
1 dried red pepper (the larger the pepper, the hotter the soup)
½ teaspoon fennel seeds
1 3-inch cinnamon stick, broken in half
6 allspice berries
2 cardamom pods
4 slices fresh gingerroot, peeled, about ⅛ inch thick
2 cups freshly squeezed orange juice
mint leaves for garnish
orange slices for garnish

• Melt the butter over medium heat in a 3-quart nonreactive pot. Add the onion and sauté until the edges are golden (about 10 minutes). Add the carrots, garlic, broth, and turmeric.
• In a spice bag or piece of cheesecloth, tie the cloves, coriander seeds, peppercorns, cumin seeds, dried red pepper, fennel seeds, cinnamon stick, allspice berries, cardamom pods, and ginger. Add the spice bag to the pot and bring to a boil. Lower the heat and simmer, uncovered, for 30 minutes, stirring occasionally.
• Remove the spice bag and puree the soup in 2 batches in a food processor or blender. Wipe out the pot and return the pureed soup to it. Add the orange juice and heat through.
• Serve hot, garnished with mint leaves and orange slices. (This soup may be made a day in advance if desired and reheated.)

Serves 6.
Preparation time: 50 minutes.

Salad of Beets, Mache, and Clementines

Salad

4 beets, 2 inches in diameter
4 clementines, peeled, sectioned, pith removed
4 handfuls mache, washed and roots trimmed
2 scallions, thinly sliced

Vinaigrette

1 tablespoon balsamic vinegar
1 tablespoon lemon juice
1 teaspoon Dijon mustard
salt and freshly ground black pepper
¼ cup good quality virgin olive oil

• Trim the beet stems and roots, leaving 2 inches of each attached to the beets. Drop them in boiling, salted water and boil for 25 to 30 minutes, or until just tender. Then plunge them into a cold-water bath until cool. Cut off the ends and slice crosswise in ⅛-inch-thick slices.
• Arrange the beet slices and clementine sections around 4 serving plates. Pile the mache in the center of each plate. Scatter the scallions over each salad.
• In a small bowl, combine the vinegar, lemon juice, mustard, and salt and pepper to taste. Whisk, adding the oil in a stream. When all the oil has been incorporated, drizzle the vinaigrette over the salads and serve.

Serves 4.
Preparation time: 45 minutes.

Ugli Fruit Caribbean Fish Salad

Salad

1½ pounds fillet of cod (or other mild white fish)
⅓ cup freshly squeezed lemon juice
1 medium green pepper, seeded and cut in 1-inch squares
3 medium ripe tomatoes, cut into ½-inch wedges
1 large cucumber, peeled, seeded, and sliced ¼ inch thick
2 bananas, peeled and sliced ¼ inch thick
1 large ugli fruit, peeled, pith removed, sectioned, and seeded
½ cup freshly grated coconut
lettuce leaves (optional)

Dressing

2 tablespoons freshly squeezed lime juice
1 cup plain yogurt or sour cream
1 garlic clove, minced
1 medium jalapeño pepper (or other green chili), seeded and minced
¼ teaspoon ground cumin
½ teaspoon freshly ground black pepper
salt to taste

• Preheat oven to 425°.
• Place the fish in a nonreactive 13- × 9-inch baking pan, slightly overlapping the pieces if necessary. Pour the lemon juice over the fish. Cover the pan with foil and bake for 8 to 10 minutes, or until the fish is just cooked through. Uncover the pan and allow the fish to cool.
• Remove the fillets to a serving bowl. Add the remaining salad ingredients. Toss gently and set aside. The fillets will break easily into bite-size pieces.
• In a small bowl, combine the dressing ingredients and mix thoroughly.
• Serve salad on a bed of lettuce, if desired; top with dressing.

Serves 6 to 8.
Preparation time: 1 hour.

Chicken Salad with Tangerines and Red Onions

Salad

3 tangerines
2 cups cooked chicken, coarsely chopped
½ cup chopped celery
1 small red onion, chopped
¼ cup chopped parsley

Dressing

3 ounces cream cheese, at room temperature
½ cup mayonnaise
1 tablespoon prepared horseradish
¼ cup milk
salt and freshly ground black pepper

• Peel the tangerines. Scrape the pith from them and finely chop enough peel to make 2 tablespoons. Reserve.
• Separate the tangerine sections and remove any pith or seeds. Halve each section and place in a bowl with the chicken, celery, onion, and parsley.
• To make the dressing, combine all the ingredients in the bowl of a food processor or blender, and pulse on and off a few seconds until thoroughly combined.
• Gently toss the chicken mixture with the dressing. Chill until ready to serve. Serve on any kind of lettuce or whole-grain bread. Garnish with reserved peel.

Serves 4.
Preparation time: 30 minutes.

Pickled Citrus Shrimp

⅓ cup extra-virgin olive oil
⅓ cup freshly squeezed lemon juice
⅓ cup freshly squeezed lime juice
3 tablespoons orange blossom honey
2 tablespoons capers, drained
1 teaspoon celery seed
1 tablespoon prepared horseradish
1 teaspoon tabasco sauce
½ teaspoon salt
1 pound large shrimp, shelled and deveined
1 large orange
1 small grapefruit
1 small red onion, halved and thinly sliced
lettuce leaves or arugula (optional)

• Combine the first nine ingredients in a large, nonreactive bowl. Whisk thoroughly to combine. Blanch the shrimp for 1 minute in boiling water. Drain the shrimp, add them to the bowl, and toss to coat.
• Peel the orange and the grapefruit and scrape off the white pith. Separate the sections, making a small slit in any sections that contain seeds, and pop them out. Add the sections to the bowl, along with the onion, and toss to coat. Cover and refrigerate overnight, tossing occasionally.
• Serve on beds of lettuce leaves or arugula, if desired. Drizzle with some of the marinade.

Serves 6 as an appetizer.
Preparation time: 20 minutes to assemble, plus 1 day to marinate.

Three · Main Courses

Lemon Fettucine with Peppered Shrimp

1 pound fresh Lemon Fettucine (recipe follows)
1 tablespoon salt
⅔ cup olive oil
¼ cup unsalted butter
1 large garlic clove, minced
1 pound (about 16) large shrimp, shelled and deveined
1 large red bell pepper, finely chopped
1 small jalapeño pepper, finely chopped
⅓ cup chopped parsley leaves
salt and freshly ground black pepper
lemon slices for garnish

• Bring a large pot of water to a boil. Add 1 tablespoon of salt, 2 tablespoons of the oil and the fettucine to the water. Stir the fettucine gently until the water returns to a boil. Boil for 1½ to 2 minutes, or until al dente. Drain.
• Meanwhile, as the water heats to boil the pasta, put the remaining olive oil and the butter in a 12-inch skillet and heat over medium-high heat. When the fettucine returns to a boil, add the garlic to the skillet and sauté for a few seconds. Add the shrimp and sauté for 1 minute on each side. Add the red and jalapeño peppers and sauté for 2 more minutes, or until shrimp is opaque. Remove from the heat and add the parsley.
• Arrange the fettucine on plates. Divide the shrimp and place them on the pasta. Spoon the sauce over each serving and add salt to taste and a generous grinding of black pepper. Garnish with the lemon slices.

Lemon Fettucine

1 cup semolina flour
1 cup unbleached, all-purpose flour
3 tablespoons finely grated lemon zest
¼ teaspoon salt
1 large egg
1 large egg yolk
1 tablespoon olive oil
2 tablespoons strained lemon juice

• If you are using an electric pasta extrusion machine, follow manufacturer's instructions.
• If you are using a manual machine, place all the ingredients in a food processor. Process until well blended. If mixture is too dry, add additional lemon juice 1 teaspoon at a time, processing for 5 seconds after each addition.
• Transfer the mixture to your work surface and knead for a few seconds until it is fairly smooth, dusting it with flour if it sticks. Wrap the mixture in plastic wrap and set it aside for 30 minutes.
• Process through the machine following manufacturer's directions.
Yield: 1 pound

Serves 4.
Preparation time: 25 minutes, plus 30 minutes for pasta to set.

Fried Scallops with Lime and Garlic

3 tablespoons freshly squeezed lime juice
1 tablespoon grated lime zest
2 teaspoons dijon mustard
2 tablespoons chopped fresh coriander (cilantro)
2 garlic cloves, peeled and crushed
1 cup extra virgin olive oil
1½ pounds sea scallops
vegetable oil for frying
¾ cup unbleached, all-purpose flour
¼ cup chili powder
2 eggs, lightly beaten
salt and freshly ground black pepper

• Combine the lime juice and zest, mustard, cilantro, and garlic in the bowl of a food processor or blender. Process a few seconds to combine; then, with motor running, add the olive oil in a stream and process until thickened and incorporated. Refrigerate until ready to serve.

• Wash the scallops and lay on paper towels to dry.

• Heat ½ inch of vegetable oil in a large skillet over medium-high heat.

• In a small bowl, combine the flour, chili powder, and salt and pepper to taste. Dredge the scallops in the mixture, rolling to coat. Shake off any excess and dip in eggs, coating them. Return to flour mixture and roll until coated. Again, shake off excess.

• Fry the scallops in several batches, being careful not to crowd them. Turn them over as they become brown and crisp. When the second side is done, remove them and drain on paper towels. Don't overcook.

• Serve hot with individual bowls of the lime sauce for dipping.

Serves 4 to 6.
Preparation time: 20 minutes.

Grilled Shark Steaks with Citrus Salsa

Citrus Salsa

2 medium oranges
1 small red onion, chopped
1 garlic clove, peeled and crushed
1 Anaheim or Coblano chili, roasted, peeled, seeded and coarsely chopped
1 tablespoon olive oil
3 tablespoons freshly squeezed lime juice
2 tablespoons chopped coriander (cilantro) leaves
1 jalapeño pepper, seeded and chopped (optional)
salt and freshly ground black pepper

Shark Steaks

4 ¾-inch-thick shark steaks, ⅓ to ½ pound each
4 teaspoons olive oil
salt and freshly ground black pepper

• Peel the oranges over a small bowl with a sharp paring knife, cutting through the peel and just barely into the flesh, then spiralling down and around from 1 end to the other, as if peeling an apple. Cut into the flesh next to each dividing membrane and pop out each section, removing any seeds. Reserve.
• In the bowl of a food processor or blender, process the remaining salsa ingredients, pulsing on and off and scraping down the sides of the bowl until well chopped. Add the oranges and continue to process until you have a chunky, well-combined mixture. Cover and let sit at room temperature for 1 hour.
• Brush the shark steaks on both sides with oil, and sprinkle with salt and pepper to taste. Grill about 3 inches from the flame for 4 to 5 minutes on each side, or until the fish feels firm but bouncy. Serve immediately with the salsa.

Serves 4.
Preparation time: 45 minutes, plus 1 hour to set.

Baked Fish with Ugli Fruit

2 tablespoons unsalted butter, softened
1 pound mild white fish fillets, such as flounder or scrod
¾-inch piece fresh gingerroot, peeled
1 small onion, thinly sliced
1 medium ugli fruit
2 teaspoons chopped fresh thyme leaves (or ¾ teaspoon dried)
salt and freshly ground black pepper

• Preheat oven to 450° and place a rack in the upper third of the oven. Grease a shallow, nonreactive pan just large enough to hold the fillets with ½ tablespoon of butter.
• Arrange the fish in a single layer in the pan. Slice the ginger paper thin and scatter it, along with the onion slices, over the fish.
• Peel the ugli fruit, removing the white pith. Separate the sections and scatter them in the pan. Top with the thyme, and salt and pepper to taste. Dot with the remaining butter and bake about 10 minutes, or until the fish is opaque and flaky.
• Pour the pan juices over the fillets when served.

Serves 4.
Preparation time: 20 minutes.

White Grapefruit-Marinated Cornish Hens

3 Rock Cornish game hens, halved and backbones removed
¼ cup mint jelly
1 cup freshly squeezed white grapefruit juice
½ teaspoon freshly ground black pepper
½ teaspoon ground cinnamon
1 tablespoon Dijon mustard
1 garlic clove, minced
grapefruit slices for garnish
mint leaves for garnish

• Place the hen halves on a countertop, bone side down. With the heel of your hand, press on each breast, flattening it somewhat. Tuck the tip of the wing under the breast. With a sharp knife, make a small slit in the skin between the breast and thigh. Stick the end of the leg through the slit. Then, place the hens in a nonreactive pan or bowl just large enough to hold them.

• In a small bowl, whisk together the remaining ingredients, except the grapefruit slices and mint leaves, until well combined. Pour the marinade over the hens; cover and refrigerate for about 4 hours.

• To cook over charcoal, grill the hens skin side down for 15 to 20 minutes. Turn and grill another 15 minutes, until the hens are tender, basting occasionally.

• To oven broil, put the hens in the broiler 4 inches from the flame under moderately high heat for 15 to 20 minutes on each side, basting occasionally with the marinade.

• Garnish with grapefruit slices and mint leaves before serving.

Serves 6.
Preparation time: 1 hour, plus 4 hours marinating time.

Chicken with Lemon and Olives

¼ cup olive oil
1½ pounds skinless, boneless chicken breasts (or thighs)
flour for dredging
2 medium onions, chopped
2 garlic cloves, chopped
1½ teaspoons chopped fresh rosemary leaves (or ½ teaspoon dried)
2 teaspoons grated lemon zest
⅓ cup freshly squeezed lemon juice
½ cup dry white wine or chicken broth
1 cup green salad olives with pimentos, drained and rinsed
freshly ground black pepper to taste

• Heat the oil in a 12-inch nonreactive skillet over medium-high heat. Dredge the chicken pieces in the flour, shaking off the excess. Brown the chicken in the oil, about 4 to 5 minutes per side. Remove the chicken to a plate and set aside.
• Sauté the onion and garlic in the remaining oil until translucent. Add the remaining ingredients and bring to a boil, scraping up any browned bits from the pan. When the sauce begins to boil, return the chicken to the pan and work the pieces down into the sauce, spooning some over each piece. Lower the heat to medium and cook 2 minutes on each side for breasts or 4 to 5 minutes for thighs.
• Serve over pasta or rice, if desired.

Serves 4 to 6.
Preparation time: 35 minutes.

Soba with Fennel, Chèvre, and Kumquats

1 medium fennel bulb, with leaves
⅓ pound kumquats, well washed
¼ cup olive oil
1 medium onion, thinly sliced
1 garlic clove, finely chopped
1 teaspoon chopped fresh thyme leaves (or ½ teaspoon dried)
1 cup dry white wine or chicken broth
salt and freshly ground black pepper
8 ounces soba (Japanese buckwheat noodles)
4 ounces goat cheese

• Bring a large pot of water to boil for the soba.
• Meanwhile, trim the fennel bulb, reserving enough of the feathery leaves to make 2 tablespoons, chopped. Quarter the bulb and cut out most of the core. Thinly slice each piece crosswise.
• Thinly slice the kumquats and remove the seeds.
• Heat the oil in a nonreactive skillet over medium-high heat. Add the onion, garlic, and sliced fennel and sauté 2 to 3 minutes. Add the thyme, kumquats, and wine or broth. Cover, lower the heat to medium, and cook 5 minutes.
• Remove from the heat and add salt and pepper to taste.
• Boil the soba al dente. Drain and arrange on 4 plates. Spoon the sauce over the noodles. Top with crumbled goat cheese and the reserved chopped fennel leaves.

Serves 4.
Preparation time: 30 minutes.

Limed Spareribs

3–4 pounds lean pork spareribs, trimmed of excess fat
¾ cup freshly squeezed lime juice
1 or 2 large jalapeño peppers, seeded and coarsely chopped
4 garlic cloves, peeled and crushed
1½ teaspoons fresh marjoram leaves (or ½ teaspoon dried)
¼ cup vegetable oil

• Place the ribs in a shallow, nonreactive pan that's just large enough to hold them. Put the other ingredients in a blender or food processor and puree. Pour the marinade over the ribs. Cover and marinate for several hours or overnight, turning the ribs once or twice.
• Preheat the oven to 350°.
• Place the ribs on a rack in a shallow roasting pan and bake for 1½ to 2 hours, basting with the marinade every 20 minutes. Turn the ribs after the first hour. When they're thoroughly browned and crisp, cut them apart, if necessary, and serve.

Serves 4.
Preparation time: 30 minutes, plus marinating time, plus 2 hours.

Pork Medallions with Mandarin Oranges and Cranberries

3 mandarin oranges
4 loin pork chops, 1½ inches thick, boned
2 tablespoons unsalted butter
2 tablespoons olive oil
salt and freshly ground black pepper
2 garlic cloves, minced
1½ teaspoons chopped fresh rosemary leaves (or ½ teaspoon dried)
½ teaspoon ground cinnamon
⅛ teaspoon ground cloves
2 tablespoons brown sugar
½ cup Marsala wine
½ cup cranberries
rosemary sprigs for garnish

• Peel the oranges. Scrape the pith from the peel and chop enough peel to make 2 tablespoons. Reserve.
• Remove any pith from the orange sections; separate the sections and remove any seeds. Reserve.
• Preheat oven to 200°.
• Pound chops on both sides with a mallet or the side of a cleaver. Re-form the chops and tie into rounds using kitchen string.
• Heat 1 tablespoon butter with the olive oil in a skillet over medium-high heat. Season pork with salt and pepper to taste and sauté the medallions for 10 to 12 minutes on each side, until well browned and cooked through. Remove the strings, place the medallions on a warm serving plate, and keep them warm in the oven until ready to serve.
• Melt the remaining butter in the pan and add the garlic, rosemary, cinnamon, cloves, brown sugar, and reserved chopped peel. Sauté, stirring a few seconds, then add the Marsala and bring to a boil. Boil until the sauce is reduced and thickened. Reduce heat to medium. Add the cranberries and cover the pan for 1 to 2 minutes, until the cranberries have burst. Uncover the pan and add the orange segments and salt and pepper to taste. Heat thoroughly and spoon the sauce over the medallions. Garnish with rosemary sprigs and serve.

Serves 4.
Preparation time: 45 minutes.

Veal Chops with Blood Oranges

½ cup unsalted butter
4 veal loin chops, about 1½ inches thick
flour for dredging
2 garlic cloves, minced
1 tablespoon chopped fresh marjoram leaves (or 1 teaspoon dried)
1½ cups blood orange juice (about 7 or 8 oranges)
sprigs of marjoram for garnish
orange zest for garnish

• Melt 6 tablespoons of butter in a 12-inch nonreactive skillet over medium-high heat.
• Dredge the chops in flour, shaking off the excess, and place in the pan. Brown for 5 minutes, turn over, and brown 5 minutes more.
• Lower the heat to medium-low and pour off the grease. Add the garlic, chopped marjoram, and juice. Cover and cook approximately 5 minutes. Uncover, turn the chops over, then re-cover and cook 5 minutes more.
• Remove the chops from the pan and keep them warm. Raise the heat, add the remaining 2 tablespoons of butter, and stir until it is incorporated into the sauce.
• Pour over the chops and serve, garnished with marjoram sprigs and orange zest.

Serves 4.
Preparation time: 25 minutes.

Beef with Tangerines

2 pounds lean beef chuck, rump, or round, cut in 1¼-inch chunks
flour for dredging
3 tablespoons vegetable oil
2 cups dry red wine
1 cup rich beef stock
¾ pound small white onions, peeled
2 large carrots, peeled and cut in 1-inch chunks
2 large garlic cloves, peeled and crushed
3 tablespoons soy sauce
½–¾ teaspoon crushed red pepper, to taste
1 teaspoon granulated sugar
5 slices fresh gingerroot, peeled, about ⅛ inch thick
2 tangerines
1 large green pepper, seeded and cut in 1-inch squares
1 large red pepper, seeded and cut in 1-inch squares
⅓ cup thinly sliced scallion greens
cooked rice (optional)

• Dredge the meat chunks in flour, shaking off any excess. Brown the meat in batches in the oil over high heat in a Dutch oven or heavy 4-quart pot. Add more oil if needed.
• Return meat to the pot; add wine, stock, onions, carrots, garlic, soy sauce, crushed red pepper, sugar, and ginger. Bring to a boil, then lower heat, cover, and simmer for 1 hour.
• Near the end of the simmering hour, remove the skin from the tangerines in as large pieces as possible, reserving the fruit. Lay the peel, outer side down, on a hard surface and, using the edge of a sharp knife or grapefruit spoon, scrape off as much white pith as possible. Cut the scraped peel into ½-inch squares and add them to the pot. Simmer for 30 more minutes.
• Add the peppers and simmer for another 15 minutes.
• Separate the reserved tangerine sections and peel off as much white pith as possible. Make a small slit in the inner edge of any section with seeds and pop the seeds out through the slit. Add the tangerine sections to the pot and heat through for 5 minutes.
• Serve, sprinkled with scallions, over rice, if desired.

Serves 6.
Preparation time: 2 hours 45 minutes.

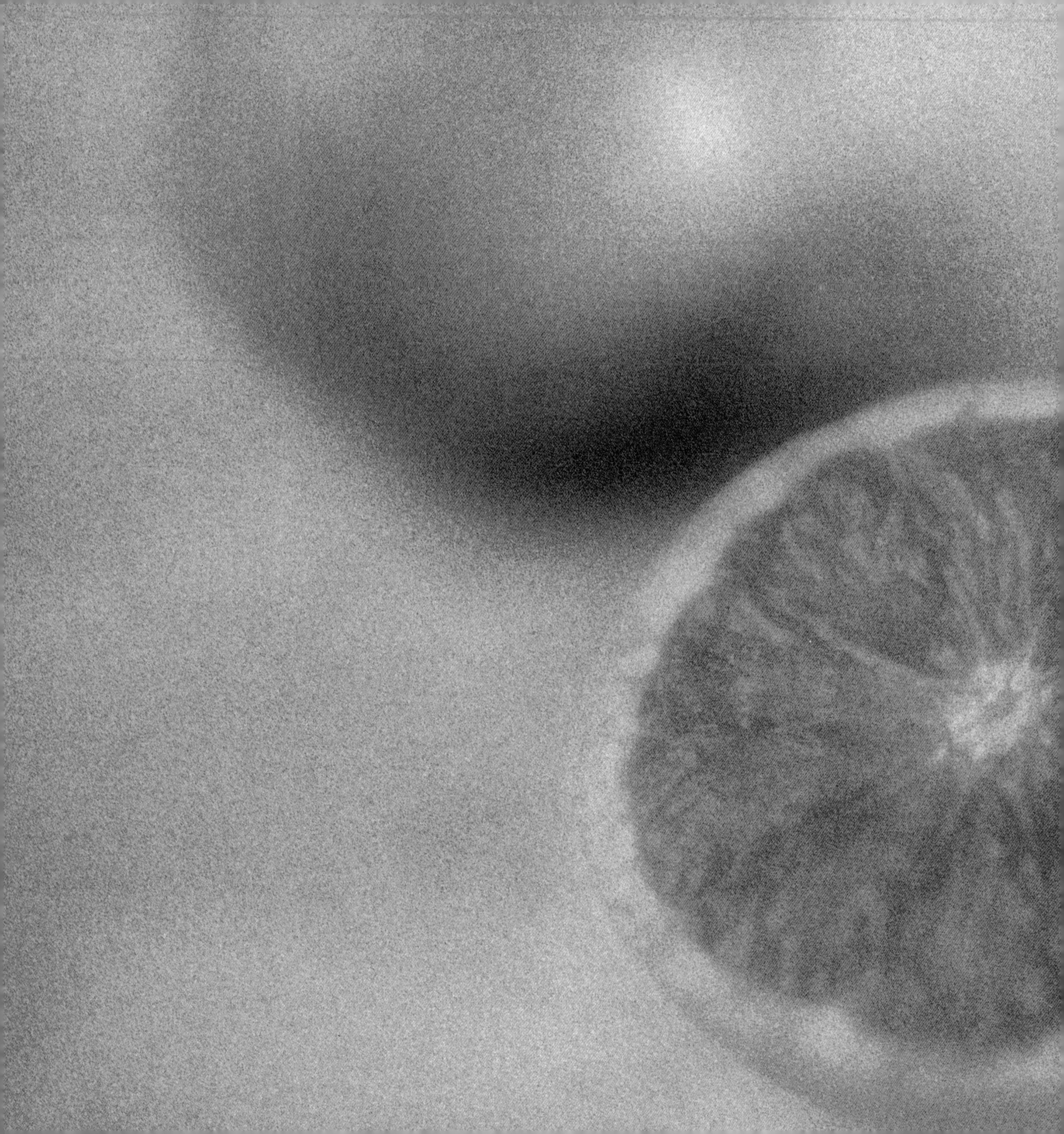

Four · Relishes and Condiments

Pickled Lemons

10–12 lemons
6 garlic cloves, crushed
6 slices fresh gingerroot, peeled, about ⅛-inch thick
1 teaspoon ground turmeric
2 teaspoons cumin seeds
2 tablespoons yellow mustard seeds
1–2 teaspoons dried red pepper flakes, to taste
½ teaspoon fennel seeds
1 teaspoon fenugreek seeds
4 cups distilled white vinegar
3 tablespoons salt

• Wash the lemons well with a scrub brush, using warm soapy water. Rinse thoroughly. Quarter them and place them, interspersed with the garlic cloves and ginger slices, in a ½-gallon widemouthed jar or pickle crock, leaving 1 to 2 inches of space at the top of the jar.
• Heat a nonreactive skillet over medium-low heat and add the turmeric, cumin seeds, mustard seeds, dried red pepper flakes, fennel seeds, and fenugreek seeds. Heat the mixture carefully, shaking the pan frequently, until you begin to smell the spices and they begin to smoke. Remove them from the heat and add 1 cup of vinegar to the pan. Mix thoroughly and carefully pour the vinegar and spices over the lemons in the jar. Add the salt and enough additional vinegar to cover the lemons.
• Seal the jar, shake it, and roll it around to mix the ingredients. Let it sit in a cool, dark place for 3 weeks, shaking the jar and rolling it around 2 or 3 times a week.
• When 2 or 3 weeks have passed, chill the lemons. They will keep in the refrigerator for 2 to 3 months.
• Slice and serve as a pickle, or chop and use as a relish with curries, meats, or greens.

Makes ½-gallon jar.
Preparation time: 30 minutes to assemble, plus 3 weeks to pickle.

Citrus Marmalade

1 small, unblemished grapefruit
1 large, unblemished orange
1 large, unblemished lemon
1 large, unblemished lime
4 cups granulated sugar
2 tablespoons Scotch whiskey (optional)

• Wash the fruit well with a scrub brush, using warm soapy water if they're not organic. Rinse thoroughly. Slice some peel from each end of each fruit, just deep enough to reach the fruit. With a sharp knife, cut the grapefruit and orange into eighths and the lemon and lime into quarters lengthwise. Lay the wedges on a flat surface with 1 cut side down and the peel side toward you. Slice the wedges as thinly as possible so that you end up with little triangles. Try to keep as much of the juice as possible in the fruit.
• Place the fruit slices into a 3½- or 4-quart nonreactive saucepan with 8 cups of water. Bring to a boil and boil rapidly for 35 to 40 minutes, or until the peel is tender, occasionally stirring down any pulp pushed up the sides of the pan. Cover and let sit overnight.
• The following day, add the sugar to the pan and heat over medium-low heat, stirring until the sugar is dissolved. Raise the heat and boil rapidly, stirring frequently, for 15 to 20 minutes, or until thickened, so that the syrup jells after cooling several minutes on the spoon or after 1 minute in the freezer. Be careful that the mixture does not stick to the pan. Remove from heat and stir in whiskey, if desired.
• Let the marmalade sit for 5 minutes, allowing it to thicken enough to maintain an even distribution of the pieces of peel. Skim off any foam, stir, and spoon into sterile jars. Seal and process in a boiling-water bath (directions follow) for 10 minutes, or cool and store in the refrigerator for 2 to 3 weeks.

Makes about 3 pints.
Preparation time: 2 hours 20 minutes, over 2 days, plus 2 or 3 weeks to stand.

Boiling-Water Bath

A boiling-water bath is not suitable for canning all types of food, but it works well for finishing and sealing the preserved foods in this chapter. Its purpose is to kill yeasts, molds, and bacteria that cannot live at 212°, the temperature of boiling water. It also forces out any air trapped in the tissues of the food and in the jar itself. Thus, it creates a vacuum that allows the jars to seal themselves for longer storage.

I find it easiest to use half-pint jars because the size of the pot increases with the jar's size. For half-pint jars you will need a 16-quart stock pot and a rack, which will allow water to circulate and hold the jars ¾ inch to 1 inch above the bottom of the pot. The same pot can be used to sterilize the lids.
• Place the jars on the rack with at least 1 inch between them. Fill the pot with water to 1½ inches above jars.
• Start timing and boil the jars 10 minutes. Remove jars with tongs and turn off heat. Fill jars with the hot preserves. Seal with new lids and return them to the pot of hot water with at least 1 inch between them. They should be covered by 1 to 2 inches of water. Add more boiling water if necessary. There should be a 1- to 2-inch space between the water and the lid for "boiling room." Turn heat back on high. Bring to a full, rolling boil. Then start timing. Process 10 to 15 minutes at a full boil.
• Remove jars with tongs and allow to cool.

Kumquat and Ginger Preserve

2 pounds kumquats
1 tablespoon baking soda
3 cups granulated sugar
1½ tablespoons fresh gingerroot, peeled and grated

• Wash the kumquats in warm soapy water and rinse thoroughly. Remove the stem ends with your fingernail or a small knife.
• Place the kumquats in a 3½-quart nonreactive pot. Sprinkle with baking soda and cover with boiling water. Let sit for 10 minutes, then drain and rinse the pot and the kumquats very thoroughly.
• Using a sharp knife, cut an "X" about ½ inch deep on the stem end of each kumquat. Return kumquats to the pot. Add fresh water to a level about 1 inch above the fruit. Boil the fruit for 15 minutes over high heat, then drain and return to the pot. Add more fresh water and boil an additional 15 minutes. Then drain, and repeat the process a third time. Drain again, and set aside.
• Place the sugar in the pot with 3 cups of fresh water. Bring to a boil and cook at a full boil for 5 minutes. Add the kumquats and the ginger. Return to a boil and boil about 30 minutes or until the kumquats are becoming translucent. (Keep the heat as high as possible without boiling over. Bubbles will rise to the top of the pot, so don't leave it unattended. Scrape any ginger that adheres to the side of the pan back into the liquid.)
• Remove the pot from the heat, cover, and let stand overnight so the kumquats can plump.
• The following day, return the pot to a boil and boil about 5 minutes, until the syrup is as thick as honey and a small amount of syrup in a spoon jells after 1 minute in the freezer.
• Pack the kumquats into sterile jars and pour the hot syrup over them, leaving ½ inch of headroom. Wipe the rims clean and seal the jars.
• If you plan to use the preserves within 2 weeks, cool and refrigerate. If you'll be storing them longer, process in a boiling-water bath (see note on page 74) for 10 minutes and store in a cool, dark place.

Makes 2 pints.
Preparation time: 2 hours 30 minutes, over 2 days.

Lime and Tomato Relish

2 large, thin-skinned limes, thinly sliced and seeded, with each slice quartered
4 large, ripe tomatoes, cored and chopped
1 large onion, peeled and chopped
1 medium green pepper, seeded and chopped
1 small hot pepper, finely chopped
1 garlic clove, finely chopped
1½ teaspoons mustard seeds
½ teaspoon celery seeds
¼ teaspoon ground cloves
1 teaspoon salt
⅓ cup granulated sugar
1 cup white vinegar

• Combine all the ingredients in a nonreactive 3-quart pot. Bring to a full boil and remove from the heat.
• Seal in hot, sterile jars. If storing, process in a boiling-water bath (see note on page 74) for 10 minutes and store in a cool, dark place.
• Tastes best if allowed to stand a week or 2 before serving. Serve with seafood or grilled meats.

Makes 4 pints.
Preparation time: 45 minutes, plus 1 or 2 weeks to stand.

Tangelo and Sweet Onion Relish

2 large, unblemished tangelos
2 large Spanish or Bermuda onions, peeled and chopped
1½ teaspoons fennel seeds
½ teaspoon crushed red pepper (optional)
½ teaspoon ground cloves
¼ teaspoon ground turmeric
½ teaspoon ground white pepper
½ teaspoon salt

• Use a vegetable peeler, pulling toward you, to remove the zest from the tangelos in strips. Stack the strips and cut them crosswise, ⅛-inch thick. Reserve.
• Using a sharp paring knife, hold 1 tangelo over a 2-quart nonreactive saucepan and cut through the white pith just into the flesh. Remove the pith in a long strip, spiralling down and around from 1 end to the other, as if peeling an apple. Cut between the membranes with the paring knife and pop the fruit sections out into the pan, making sure to remove any seeds. Squeeze the translucent membranes over the pan to extract any remaining juice. Repeat the process with the second tangelo.
• Add the onion, the spices, the reserved zest, and ½ cup water to the pan and cook at medium heat until the mixture comes to a boil. Then, lower the heat and simmer until the onion begins to look translucent (about 5 minutes).
• Store in the refrigerator for up to 2 weeks in airtight containers. Or, seal in sterile jars and process in a boiling-water bath (see note on page 74) for 10 minutes and store in a cool, dark place.
• Serve with hamburgers, roasted or grilled meats, or stir into rice pilaf.

Makes about 2 pints.
Preparation time: 45 minutes.

Tangerine-Cranberry Chutney

4 tangerines
12 ounces fresh cranberries
2 cups granulated sugar
½ cup distilled white vinegar
½ cup golden raisins
2 tablespoons fresh gingerroot, peeled and chopped
1 garlic clove, minced
1 red hot pepper, seeded and finely chopped (the larger, the hotter)
1 3-inch cinnamon stick
6 whole cloves
4 whole allspice berries

• Peel the tangerines. Scrape the white pith strands from them and sliver enough peel to make ¼ cup.
• Place the peel in a large nonreactive saucepan with the cranberries, sugar, vinegar, raisins, ginger, garlic, and hot pepper. Tie the spices in a spice bag or piece of cheesecloth and add to the pot.
• Cook over medium heat, stirring constantly until the sugar is dissolved and the cranberrries begin to pop. Simmer the mixture for 8 to 10 additional minutes, stirring frequently.
• Meanwhile, scrape off any white pith from the tangerines. Separate the segments and cut off the inner edge of each one, removing any seeds. Chop the tangerine sections coarsely and add them to the pot.
• Raise the heat and bring the mixture to a boil, then simmer for 5 minutes, until it has thickened, stirring often to prevent sticking.
• Remove the spice bag and ladle the mixture into hot, sterile jars. Seal and process in a boiling-water bath (see note on page 74) for 10 minutes. Then store in a refrigerator or cool place for up to 3 weeks. It's best to let the chutney stand for at least a week before serving.

Makes 3 pints.
Preparation time: 45 minutes, plus at least 1 week for curing.

Orange Butter

2 oranges
1½ cups granulated sugar
2 whole eggs
3 egg yolks
pinch of salt
1 stick (8 ounces) unsalted butter

• Using a vegetable peeler, remove the zest from both oranges. Cut the zest into ½-inch pieces and place in the bowl of a food processor or blender with ½ cup sugar. Process on and off about 30 seconds, or until the zest is finely chopped. Reserve.
• Squeeze both oranges thoroughly and reserve the juice.
• Put water 2 inches deep in the bottom of a double boiler and bring to a simmer over medium heat.
• In a mixer or by hand with a whisk, beat together the eggs, egg yolks, and salt until foamy. Gradually add the remaining sugar, and then the zest mixture, and continue beating until very thick (about 2 minutes). Pour the mixture into the top of the double boiler and place it over the simmering water. Whisk constantly while adding the reserved orange juice. Continue whisking constantly until it is thick, steaming hot, and smooth, and the zest pieces no longer sink to the bottom (about 12 minutes). Add butter ½ tablespoon at a time, whisking until incorporated.
• Spoon the butter into hot, dry, sterile jars and allow to cool. Then seal and refrigerate for up to 2 months. Spread on bread or toast or between cake layers or cookies. Or, while hot, spoon into baked, individual tart shells, allow to cool, and serve chilled with fresh fruit.

Makes approximately 1 pint.
Preparation time: 30 minutes.

Five · Desserts

Vermont Orange Ambrosia

4 large oranges, peeled and pith removed
⅓ cup maple syrup
¼ cup dark rum
¼ teaspoon ground cinnamon
¾ cup large walnut pieces
½ cup pomegranate seeds (optional)

• Halve the oranges lengthwise and slice into ¼-inch-thick half-rounds. Place in a serving bowl.
• Put the maple syrup, rum, and cinnamon in a small bowl, stirring constantly until well blended. Pour over oranges, tossing gently to coat. Cover and refrigerate for several hours or overnight.
• Preheat oven to 350°. Put the walnut pieces in a small, shallow pan and toast them for 15 minutes, or until lightly browned. Cool.
• When you're ready to serve dessert, toss the nuts with the oranges, sprinkle with the pomegranate seeds, if desired, and serve immediately.

Serves 4 to 6.
Preparation time: 30 minutes, plus several hours to chill.

Chocolate-Dipped Citrus Sections

6 ounces semisweet or bittersweet chocolate
4 tablespoons unsalted butter
2 cups mixed citrus sections from easy-to-peel fruits, such as navel oranges, tangerines, ugli fruit, and clementines, or whole thin-skinned lemons cut into eighths
finely chopped pistachio nuts (optional)

• Melt the chocolate and butter in the top of a double boiler over simmering water, stirring to mix thoroughly.
• Remove any white pith from the citrus sections (except lemons), and any seeds, by making a small slit in the membranes with a sharp knife and popping the seeds out.
• Dip each section into the chocolate, coating about half of its length. Place the dipped sections on parchment paper or foil and refrigerate until the chocolate is set. If desired, sprinkle pistachios over the dipped sections while the chocolate is still moist.

Serves 6 to 8.
Preparation time: 30 minutes.

Minted Ruby Grapefruit Ice

¾ cup granulated sugar
2 cups ruby-red grapefruit juice, freshly squeezed, with pulp (4 or 5 grapefruits)
2 tablespoons finely chopped mint leaves
mint sprigs for garnish

• In a small saucepan, heat the sugar with ½ cup of water, stirring until the sugar is dissolved. Boil for 5 minutes. Cool to room temperature, then combine the sugar syrup with the grapefruit juice and the chopped mint leaves.
• Freeze the mixture in an ice cream maker, following the manufacturer's instructions, or place it in a freezerproof container and freeze for several hours or overnight. Serve garnished with mint sprigs.

Serves 6. Makes 1 pint.
Preparation time: 30 minutes, plus several hours to freeze.

Tangerine Sorbet

½ cup granulated sugar
2 tablespoons tangerine peel, with all white pith removed, finely chopped
2 cups freshly squeezed tangerine juice (about 6 to 8 tangerines)

• In a small nonreactive saucepan, combine the sugar, peel, and ½ cup water. Bring the mixture to a boil, stirring until all the sugar has dissolved. Boil for 5 minutes. Remove from heat and let cool to room temperature.
• Combine the syrup with the juice and freeze in an ice cream maker, following the manufacturer's instructions, or place in a freezerproof container and freeze for several hours or overnight before serving.

Serves 6. Makes 1 pint.
Preparation time: 30 minutes, plus several hours to freeze.

Lemon Snaps

2½ cups unbleached, all-purpose flour
1½ cups granulated sugar
2 teaspoons baking soda
¼ teaspoon salt
2 tablespoons grated lemon zest
¾ cup vegetable oil
½ cup freshly squeezed lemon juice
2 teaspoons vanilla extract

• Preheat oven to 350°.
• Place all the ingredients in a food processor or mixer and process or mix by hand until well blended. Drop by the teaspoonful onto greased cookie sheets about 2 inches apart. Bake about 10 minutes, or until the edges of the cookies are golden brown.
• Cool on racks and store in an airtight container.

Makes about 4 dozen cookies.
Preparation time: 40 minutes.

Chocolate-Orange Marmalade Brownies

1½ cups pecans or walnuts
10 tablespoons unsalted butter
5 ounces unsweetened chocolate
1 cup good-quality orange marmalade
3 tablespoons Scotch whiskey
1 teaspoon orange extract
1 cup unbleached, all-purpose flour
½ teaspoon salt
4 large eggs
1⅓ cups granulated sugar

• Preheat oven to 375°.
• Spread the nuts in a large, shallow baking pan and toast in the oven about 8 minutes, or until lightly toasted. Remove and cool, then break them into pieces and reserve.
• Grease a 13- × 9- × 1½-inch baking pan with 1 teaspoon of the butter.
• In the top of a double boiler, over simmering water, place the remaining butter and the chocolate. Cover and heat until the chocolate is almost completely melted. Remove from the heat. Uncover and stir until smooth. Add the marmalade, stirring until it has melted and combined with the chocolate. Stir in the Scotch whiskey and orange extract and set aside until it is warm, but no longer hot.
• Sift the flour and salt and reserve.
• In the bowl of a mixer or by hand, beat the eggs until frothy. Add the sugar and beat until light colored and thickened (about 5 minutes). With the mixer on low, add the warm chocolate mixture, scraping the bowl and beating the mixture for a few seconds to combine. Add the flour and salt. Scrape the bowl and beat just until the flour has been combined. Do not overbeat.
• By hand, stir in the nuts and pour the batter into the prepared pan, smoothing it out with a spatula.
• Bake about 25 minutes, or until a toothpick inserted in the center comes out clean.
• Cool in the pan. Cut into 18 3- × 2-inch brownies and serve or wrap individually in plastic wrap. Full flavor develops after 24 hours; serve alone or topped with ice cream.

Makes 18 brownies.
Preparation time: 1 hour.

Orange-Coconut Custard

2 medium oranges
1 cup milk
3 large eggs
3 large egg yolks
¼ cup granulated sugar
15-ounce can coconut cream
1 cup shredded coconut
additional zest to garnish (optional)

• Preheat oven to 350°.
• With a vegetable peeler, remove the zest from the oranges in long strips. Reserve the oranges.
• Place the zest in a small, heavy saucepan with the milk and heat over medium-low heat until it is steaming hot. Remove from the heat. Do not let the mixture boil.
• In a mixing bowl, whisk together the eggs, egg yolks, and sugar, beating until frothy. Add the coconut cream, whisking to mix well.
• Remove the zest from the milk and discard. Add the hot milk to the mixing bowl in a stream while whisking. Strain the mixture and reserve.
• Take the reserved oranges and, with a sharp paring knife, remove the white pith and cut just through the membranes and into the fruit. Spiral down and around the fruit as if peeling an apple. Insert the knife next to the dividing membranes and pop out each section. Remove the seeds and coarsely chop the fruit. Divide the pulp among each of eight 6-ounce custard cups.
• Fill the cups ¾ full with the custard mixture and sprinkle 1 tablespoon of shredded coconut on the top of each cup. Place the custard cups in a baking pan large enough to hold them. Fill with hot water to half the depth of the cups. Cover the pan with foil and bake the custard for 50 to 55 minutes or until set. Serve warm, or cool and refrigerate a couple of hours or overnight before serving. Garnish with zest if desired.

Serves 8.
Preparation time: 1 hour 30 minutes, set overnight.

Lime Mousse Pie

1 packet unflavored gelatin
3 eggs
1 cup granulated sugar
⅓ cup freshly squeezed lime juice (about 4 limes, key limes may be used)
2 teaspoons finely grated lime zest (key limes may be used)
2 tablespoons powdered sugar
1 teaspoon cornstarch
1 cup heavy cream, cold
⅛ teaspoon salt
1 9-inch piecrust, baked and completely cooled

• In a small bowl, soften the gelatin in ¼ cup cold water. Set aside.
• Separate the eggs, reserving the whites at room temperature. Place the yolks in the top of a double boiler over simmering water. Add ⅔ cup sugar and cook, stirring constantly, until thickened, about 5 minutes.
• Add the softened gelatin to the hot custard and stir well, dissolving the gelatin completely. Add the lime juice and zest and mix thoroughly.
• Cool the mixture and chill until slightly thickened, about 1 hour.
• Place the powdered sugar and the cornstarch in a small saucepan and gradually add ¼ cup cream. Bring to a boil, stirring constantly, then simmer briefly until the mixture thickens. Remove from the heat and cool to room temperature.
• When the custard has cooled and thickened, beat the egg whites with the salt until they begin to hold a soft shape. Add the remaining ⅓ cup sugar 1 tablespoon at a time, beating until the sugar is well incorporated and the whites are stiff, but not dry.
• Using the same beaters, whip the cream in a chilled bowl until the beaters leave marks. Add the cornstarch mixture in a steady stream and continue beating until the cream holds stiff peaks. Do not overbeat.
• Fold the cream into the custard and then gently fold in the egg whites. Spoon into the piecrust and refrigerate 4 hours or more before serving.

Serves 8.
Preparation time: 2 hours, plus 4 hours for cooling.

Frozen Blood Orange Soufflé

4 large blood oranges
¼ cup Grand Marnier or
orange liqueur
½ cup granulated sugar
1 envelope unflavored gelatin
3 egg whites
pinch of salt
1 cup heavy cream

• Cut six 6-inch pieces of waxed paper (make sure they are large enough to extend around the rims of six 6-ounce ramekins). Fold each piece in half lengthwise and fasten around the top edge of each ramekin with string or rubber bands. The waxed paper should extend 2 to 2½ inches above the rims.

• Grate enough zest from the oranges to make ¼ cup loosely packed. Place the zest in a small stainless steel, glass, or enamel pan, along with the Grand Marnier and the sugar. Simmer over low heat for 10 to 15 minutes, stirring frequently, until all the sugar is dissolved and the zest is translucent. Remove from heat.

• Peel the oranges with a sharp knife, carefully removing all the white pith. Cut into sections and scrape out the pulp, removing the seeds in the process. The oranges should yield 1½ to 2 cups of pulp.

• Place the pulp in the bowl of a food processor, along with the zest mixture, and puree briefly. Place the mixture in a 1-quart stainless steel, glass, or enamel pan. Sprinkle the gelatin over the mixture, and set it aside for a few minutes to soften. Then heat the mixture over medium-low heat, stirring constantly, until all of the gelatin is dissolved. Remove it from the stove, put it into a large bowl, and cover it. Place the bowl in the refrigerator, stirring occasionally, until the mixture has cooled and mounds when dropped from a spoon (about 30 minutes).

• In a small mixing bowl, beat the egg whites with the salt until they hold stiff peaks. Add the egg whites to the orange-gelatin mixture. Using the same beaters, beat the cream in another bowl until it holds stiff peaks. Do not overbeat. Add the whipped cream to the other ingredients and gently fold together until thoroughly combined. Spoon the mixture into the ramekins and freeze for 2 to 3 hours, until set. Remove the waxed paper collars and serve.

Serves 4.
Preparation time: 3 hours 30 minutes to 4 hours.

Lemon-Nutmeg Cheesecake

1½ cups vanilla wafer crumbs (about 36 wafers crushed with a rolling pin or in a food processor)
1½ cups granulated sugar
3 tablespoons finely grated lemon zest
1½ teaspoons freshly grated nutmeg
¼ cup unsalted butter, melted
2 pounds cream cheese, at room temperature
⅓ cup unbleached, all-purpose flour
4 large eggs, at room temperature
½ cup heavy cream, at room temperature
½ cup freshly squeezed lemon juice
2 teaspoons vanilla extract

• Preheat oven to 350°.
• Generously grease a 9-inch spring-form pan and place it in the center of a 12-inch-square piece of foil. Fold the foil up around the sides of the pan and press to make a tight fit.
• In a small bowl, combine the cookie crumbs with 3 tablespoons of sugar, 2 tablespoons of zest, and ½ teaspoon nutmeg. Stir to blend. Add the melted butter and mix thoroughly. Press the crumb mixture into the bottom of the pan and up the sides about 1 inch. Bake the crust for 10 minutes. Remove it from the oven and let cool for 10 to 15 minutes.
• In the large bowl of a mixer or by hand, beat the cream cheese at low speed until soft. Add the remaining sugar and beat at medium speed until light and fluffy. Add the flour and remaining nutmeg and beat until combined. Add the eggs, one at a time, beating well after each addition.
• Combine the heavy cream, lemon juice, vanilla extract, and remaining zest and fold it into the mixture until it is smooth.
• Pour the filling into the crust and bake in the center of the preheated oven for 45 minutes or until the top is golden. Turn off the heat and allow the cake to sit undisturbed in the oven for 45 minutes.
• Remove the cake from the oven and let it cool completely in the pan. Cover and chill for 4 hours or overnight. Remove the sides of the pan and allow the cake to warm slightly before serving.

Serves 8 to 10.
Preparation time: 2 hours, set 4 hours or overnight.

Lemon Steam Cake with Blood Orange Sauce

3 large eggs
1 cup granulated sugar
grated zest of 2 lemons
1½ cups unbleached, all-purpose flour
2 teaspoons baking powder
¼ teaspoon salt
1 cup heavy cream
1 teaspoon vanilla extract

• Grease and flour a 9-inch spring-form pan. Place it in the center of a 12-inch-square piece of foil and fold the foil up the sides of the pan to make the bottom watertight.
• In a large mixing bowl, beat the eggs until they are thick and pale yellow. Beat in the sugar until the mixture is smooth and well combined. Beat in the lemon zest.
• Sift the flour, baking powder, and salt together.
• Combine the cream and the vanilla.
• Add the flour mixture and the cream mixture alternately to the mixing bowl, about a third at a time, beating well after each addition.
• When it is well mixed, pour the batter into the pan. Place the pan on a rack in a wok or large pot filled with enough simmering water to almost reach the rack. (A vegetable steamer works well.) Cover tightly and steam over medium-low heat for 45 to 50 minutes, until a cake tester inserted in the center comes out clean. Remove the cake to a rack and let cool.
• When the cake has cooled completely, run a knife around the edge of the pan and carefully remove the sides. Slice and serve with Blood Orange Sauce (recipe follows).

Blood Orange Sauce

1 cup strained blood orange juice (5–6 oranges)
½ cup sugar
2 whole cloves
1 teaspoon cornstarch
1 tablespoon Grand Marnier or orange liqueur

• Combine the juice, sugar, and cloves in a small, nonreactive pan over medium heat. Stir the mixture until the sugar dissolves. Simmer for 5 minutes.
• Dissolve the cornstarch in the Grand Marnier and stir it into the sauce. Simmer 5 minutes more.
• Remove the cloves and serve with the cake, either hot, at room temperature, or chilled.

Serves 6 to 8.
Preparation time: 1 hour 15 minutes.

Mandarin Orange Upside-Down Cake

5 mandarin oranges
¼ cup Grand Marnier or orange liqueur
¾ cup unsalted butter, softened
½ cup light brown sugar
3 large eggs, separated
1 teaspoon vanilla or orange extract
1½ cups sifted cake flour (not self-rising)
¾ cup granulated sugar
¾ teaspoon baking powder
¼ teaspoon baking soda
¼ teaspoon salt
½ cup plain yogurt or sour cream
whipped cream for garnish (optional)

• Peel the oranges and separate the sections, scraping off as much of the white pith as possible. Cut off the inner edge of each section and remove any seeds. Place the sections in a small nonreactive bowl and toss gently with the Grand Marnier. Let sit for 30 minutes, tossing occasionally, and reserve.

• Preheat oven to 350°.

• Melt 4 tablespoons of the butter in a well-seasoned, 10-inch cast iron skillet over medium heat. Stir in the brown sugar until it's well moistened. Spread the sugar mixture evenly over the bottom of the skillet and remove it from the heat.

• Arrange the marinated orange sections in a spiral pattern on the bottom of the pan, starting from the center and working outward. Reserve the liquid.

• Add the egg yolks and vanilla or orange extract to the liquid remaining in the bowl. Beat lightly to combine and set aside.

• In a large mixing bowl, combine the flour, sugar, baking powder, baking soda, and salt. Mix on low speed for 20 seconds to blend. Add the yogurt or sour cream and the remaining softened butter and continue to mix on low speed until well combined. Increase speed to medium and beat for 2 minutes. Add the egg-yolk mixture, a third at a time, beating well after each addition. Set aside.

• In a small mixing bowl, using clean beaters, beat the egg whites until stiff. Fold a third of the egg whites into the batter. When well incorporated, gently fold in the remaining egg whites.

• Pour the batter carefully over the orange sections so that you do not disturb their arrangement. Make sure the batter is distributed evenly and all orange sections are covered.

• Bake in the lower third of the oven for 40 to 45 minutes. The cake is done when it springs back when touched lightly in the center or a cake tester or toothpick inserted in the center comes out clean.

• Allow to cool in the pan for 5 minutes. Then run a knife around the inside of the pan, place a serving plate over the pan, invert, and carefully lift the pan off the cake. Replace any orange sections that stick to the pan.

• Serve warm or at room temperature, with whipped cream if desired.

Serves 6 to 8.
Preparation time: 2 hours.

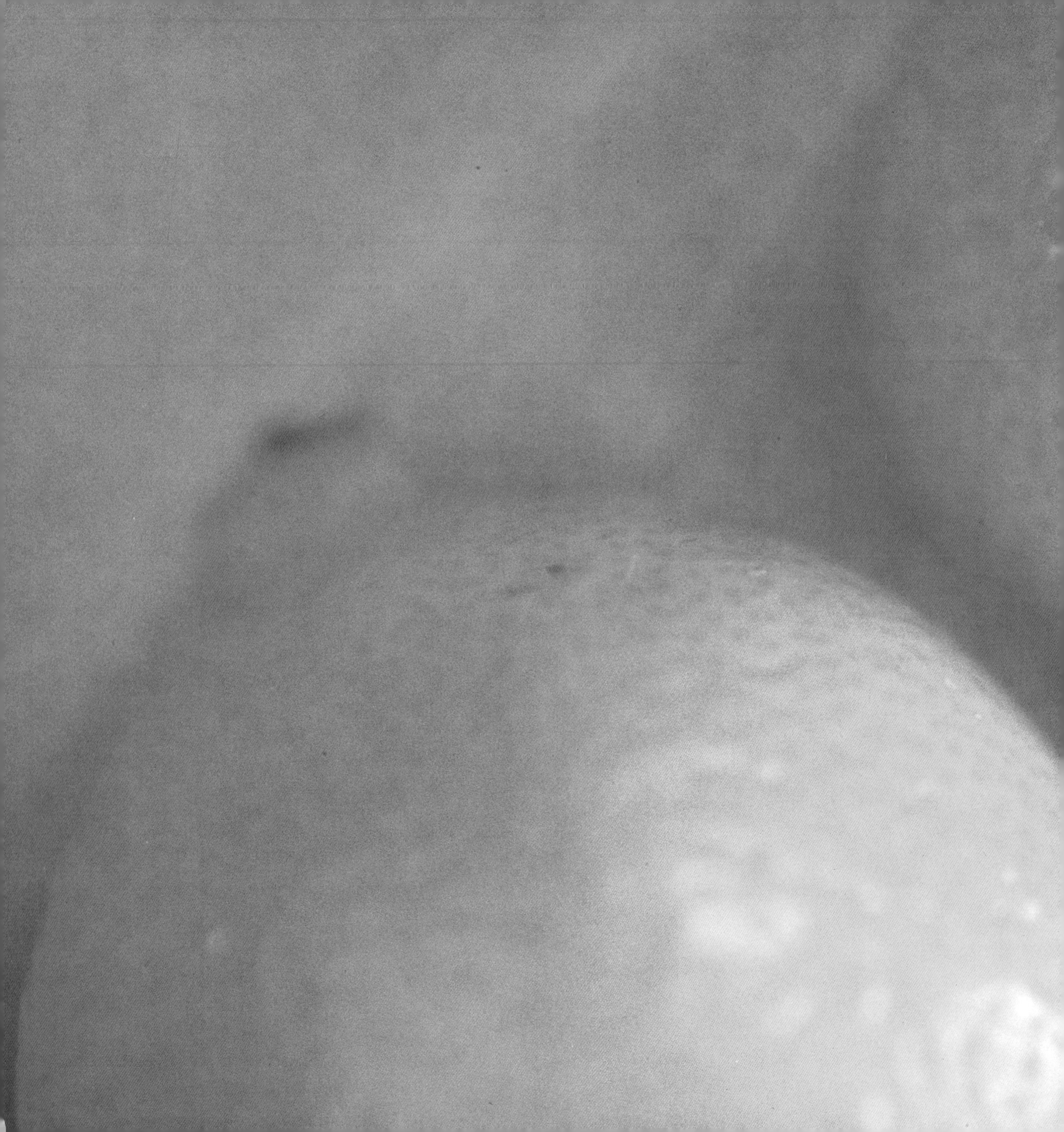

Six · Beverages

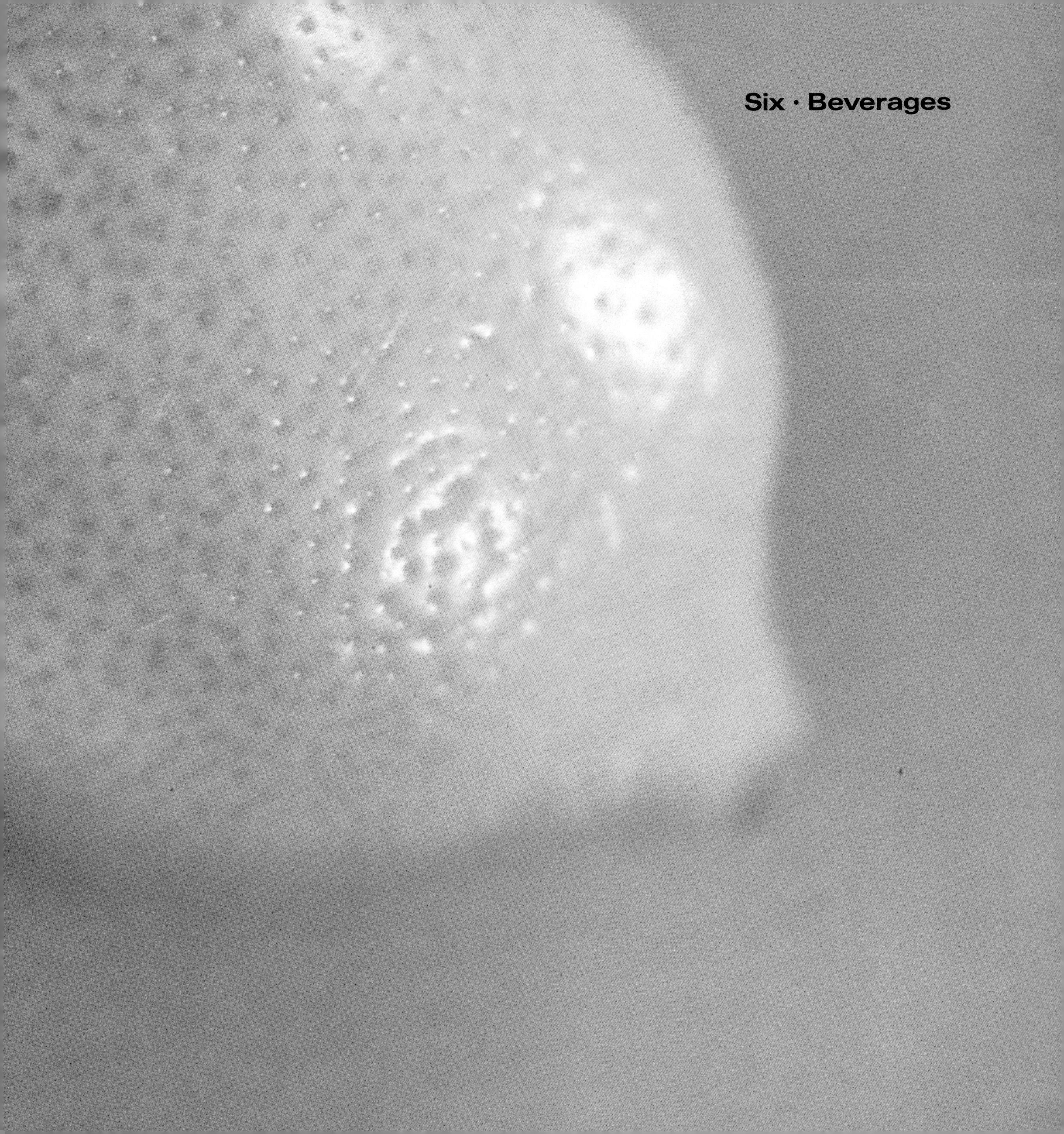

Pink Lemonade

¾ cup freshly squeezed lemon juice (4–5 lemons)
⅔ cup superfine sugar, or to taste (granulated may be substituted)
½ lemon, thinly sliced and seeded
several drops of grenadine syrup

• Combine the lemon juice and sugar in a 1-quart pitcher. Stir to dissolve the sugar.
• Add 3 cups of cold water and the lemon slices and mix well. Add drops of grenadine syrup until you achieve the desired "pinkness." Serve over ice.

Makes 1 quart.
Preparation time: 10 minutes.

Iced Orange Coffee

2½ cups strong coffee
1 small orange
cream and sugar
ground cinnamon
orange slices for garnish (optional)

• Brew the coffee.
• Using a vegetable peeler, remove the zest from the orange in strips. Put the strips in the hot coffee. Set aside for 1 hour to cool to room temperature.
• Strain the coffee.
• Squeeze the orange and add the juice to the coffee. Serve over ice with cream and sugar, as desired. Sprinkle the top of each serving with a pinch of cinnamon and garnish with an orange slice if desired.

Makes four 10-ounce servings.
Preparation time: 5 minutes, plus 1 hour to cool.

Limeade Fizz

By the glass

2 tablespoons freshly squeezed lime juice
1 tablespoon superfine sugar (granulated sugar may be substituted)
2 ounces vodka, gin, rum, or tequila
½ cup seltzer or sparkling water
crushed ice
lime slice for garnish

• In a 10-ounce glass, place the lime juice, sugar, and liquor. Stir well. Add the seltzer or sparkling water and fill the glass with crushed ice. Garnish with lime slice and serve.

Serves 1.
Preparation time: 3 minutes.

By the pitcher

⅔ cup freshly squeezed lime juice
⅓ cup superfine sugar
1¼ cups vodka, gin, rum, or tequila
2 cups seltzer or sparkling water
lime slices
crushed ice

• Combine the lime juice, sugar, and liquor in a 1-quart pitcher. Stir until the sugar is dissolved. Add the seltzer or sparkling water and the lime slices. Stir gently to combine. Pour into glasses filled with crushed ice.

Serves 4.
Preparation time: 5 minutes.

Citrus Frappé

juice of 1 large lime
juice of 1 small grapefruit
juice of 1 medium lemon
juice of 1 medium orange
½ cup superfine sugar (granulated sugar may be substituted)
3 cups crushed ice

• Combine the juices. (They should make approximately 1 cup of liquid.)
• Place the juices and the other ingredients in a food processor or blender and mix until smooth.
• The frappé will keep in the freezer for an hour or so.

Makes two 10-ounce servings.
Preparation time: 5 minutes.

Hot Spiked Lemonade

zest of ½ lemon, removed in strips
¾ cup freshly squeezed lemon juice
¾ cup granulated sugar
3-inch cinnamon stick
⅓ cup brandy, cognac, or light rum
very thin slices of lemon for garnish (optional)

• Place 3¼ cups of water and all of the ingredients, except the liquor and lemon slices, into a 1½-quart non-reactive saucepan. Bring the mixture to a boil, stirring to dissolve the sugar. Lower the heat and simmer for 5 minutes.
• Remove the pan from the heat. Take the cinnamon stick and the lemon strips out of the pan and discard. Add the liquor and serve garnished with lemon slices, if desired.

Makes four 8-ounce servings.
Preparation time: 20 minutes.

Lemon-Thyme Tea

2 tablespoons fresh thyme leaves or 1 tablespoon dried thyme
zest of 1 lemon, removed in strips
lemon wedges (optional)
honey (optional)

• Place the thyme leaves and lemon zest in a 1-quart teapot. Add 3 cups of boiling water and let steep for 5 minutes. Strain and serve with lemon wedges and honey, if desired.
• For variety, strain and add honey, stirring to dissolve it. Then, cool and refrigerate before serving over ice with lemon wedges and thyme sprigs for garnish.

Makes three 1-cup servings.
Preparation time: 10 minutes.

Sherried Pink Grapefruit Juice

2 cups freshly squeezed pink grapefruit juice
1 cup good-quality dry sherry
ice cubes
pink grapefruit slices or wedges for garnish

• Combine the juice and the sherry. Chill for an hour or more.
• Fill four 10-ounce glasses with ice cubes and the juice mixture. Garnish with grapefruit slices and serve.

Makes four 10-ounce servings.
Preparation time: 5 minutes.

White Grapefruit–Pineapple–Rum Punch

2½ cups freshly squeezed white grapefruit juice
1½ cups fresh or unsweetened canned pineapple juice
½ cup superfine sugar
1½ cups white rum
1 liter club soda or seltzer
pineapple and grapefruit slices (optional)
mint leaves (optional)

• In a bowl or pitcher, combine the juices. Add the sugar, stirring to dissolve. Add the rum and stir until well blended. Chill thoroughly.
• When ready to serve, gently stir in the club soda or seltzer and, if desired, garnish with the fruit slices and mint leaves. Serve "neat," or over ice.

Makes six 10-ounce servings.
Preparation time: 10 minutes.

Blood Orange Liqueur

zest of 1 blood orange
5 cups blood orange pulp (about 6 or 7 oranges)
1 pint vodka
1½ cups granulated sugar

• Remove the zest with a vegetable peeler in long strips. Reserve both the zest and the orange. Peel the rest of the oranges and scrape off as much of the white pith from the fruit sections as possible.
• With a sharp paring knife remove the white pith from the reserved orange, cutting just through the membrane and into the fruit. Spiral down and around the fruit as if peeling an apple. Insert the knife next to the dividing membrane and pop out each section. Place the orange sections in a food processor or blender and puree briefly.
• Place the pulp, zest, and vodka in a ½-gallon glass jar. Cover it with waxed paper and seal it with a lid, making the jar airtight. Let the mixture sit for 1 month in a cool place, shaking gently a few times a week.
• When the month is up, strain the mixture through several layers of cheesecloth spread in a strainer. When only pulp remains, pull up the ends of the cheesecloth, enclosing the pulp, and twist it to extract as much liquid as possible.
• Return the strained liquid to the ½-gallon jar. (Make sure the jar has been well rinsed.)
• Combine the sugar with ¾ cup of water in a small pan. Bring to a boil, stirring constantly, being careful not to let it boil over. When all the sugar has dissolved and the syrup is clear, remove from the heat and cool to room temperature.
• Add the room-temperature syrup to the vodka-juice mixture. Stir well. Seal again with waxed paper and lid. Let sit for 2 or 3 days, until the liqueur has cleared and sediment has settled. Syphon or carefully pour the clear portion of the liqueur into 1-pint bottles. Seal and let sit for several months, so the liqueur can mellow and age. (Use the remaining portion with the sediment for cooking, or drink it yourself.)

Makes three 1-pint bottles.
Preparation time: 30 minutes, plus several months (at least 6) for aging.

RECIPE LIST

Baked Fish with Ugli Fruit•54
Beef with Tangerines•68
Blood Orange Liqueur•116
Buckwheat Crepes with Clementines•26
Chicken Salad with Tangerines and Red Onions•42
Chicken with Lemon and Olives•58
Chocolate-Dipped Citrus Sections•86
Chocolate-Orange Marmalade Brownies•92
Citrus Frappé•110
Citrus Marmalade•74
Curried Orange Soup•36
Fried Scallops with Lime and Garlic•50
Frozen Blood Orange Soufflé•98
Grilled Pink Grapefruit and Pork Skewers•30
Grilled Shark Steaks with Citrus Salsa•52
Hot Spiked Lemonade•112
Iced Orange Coffee•108
Kumquat and Ginger Preserve •76
Lemon and Allspice Muffins•18
Lemon Fettucine with Peppered Shrimp•48
Lemon-Nutmeg Cheesecake•100
Lemon-Parsley Soup•34
Lemon Snaps•90
Lemon Steam Cake with Blood Orange Sauce•102
Lemon-Thyme Tea•112
Limeade Fizz•110
Lime and Tomato Relish•78
Limed Spareribs•62
Lime Mousse Pie•96
Lime Soup•32
Mandarin Orange Upside-Down Cake•104
Minted Ruby Grapefruit Ice•88
Navel Orange Waffles with Blueberry Sauce•22
Orange-Banana Muffins•18
Orange Butter•80
Orange-Coconut Custard•94
Orange French Toast•24
Pickled Citrus Shrimp•44
Pickled Lemons•72
Pink Grapefruit Poached in Sauternes•14
Pink Lemonade•108
Pork Medallions with Mandarin Oranges and Cranberries•64
Salad of Beets, Mache, and Clementines•38
Sherried Pink Grapefruit Juice•114
Soba with Fennel, Chèvre, and Kumquats•60
Stewed Kumquats and Strawberries•16
Tangelo and Sweet Onion Relish•78
Tangerine-Cranberry Chutney•80
Tangerine-Pecan Scones•20
Tangerine Sorbet•88
Ugli Fruit Caribbean Fish Salad•40
Veal Chops with Blood Oranges•66
Vermont Orange Ambrosia•84
White Grapefruit-Marinated Cornish Hens•56
White Grapefruit-Pineapple-Rum Punch•114

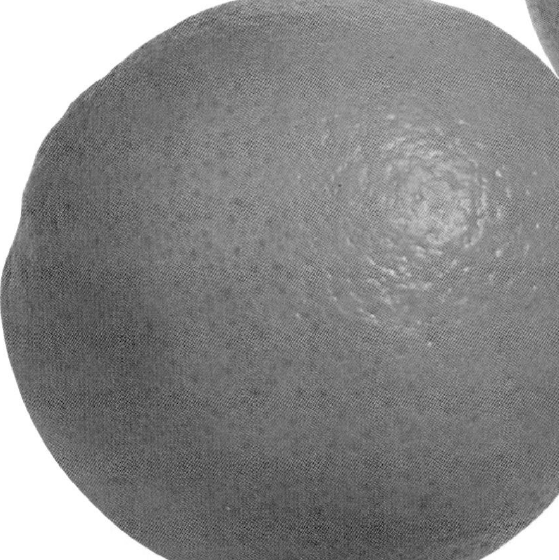

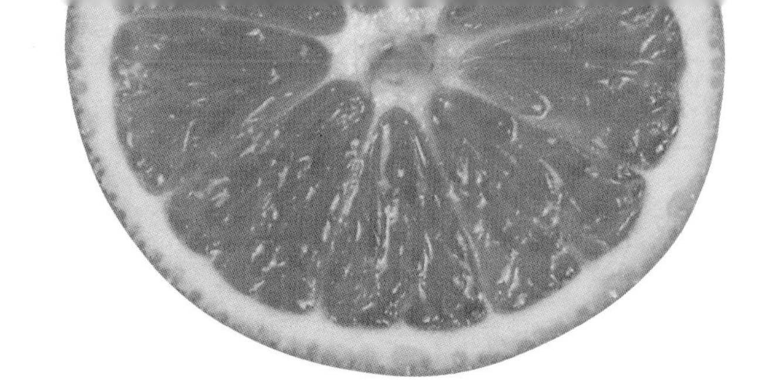

RECIPE LIST BY CITRUS

CLEMENTINES

Buckwheat Crepes with Clementines • 26
Salad of Beets, Mache, and Clementines • 38

CITRUS

Chocolate-Dipped Citrus Sections • 86
Citrus Frappé • 110
Citrus Marmalade • 74
Grilled Shark Steaks with Citrus Salsa • 52
Pickled Citrus Shrimp • 44

GRAPEFRUIT

Grilled Pink Grapefruit and Pork Skewers • 30
Minted Ruby Grapefruit Ice • 88
Pink Grapefruit Poached in Sauternes • 14
Sherried Pink Grapefruit Juice • 114
White Grapefruit-Marinated Cornish Hens • 56
White Grapefruit-Pineapple-Rum Punch • 114

KUMQUATS

Kumquat and Ginger Preserves • 76
Soba with Fennel, Chèvre, and Kumquats • 60
Stewed Kumquats and Strawberries • 16

LEMONS

Chicken with Lemon and Olives • 56
Hot Spiked Lemonade • 112
Lemon and Allspice Muffins • 18
Lemon Fettucine with Peppered Shrimp • 48
Lemon-Nutmeg Cheesecake • 100
Lemon-Parsley Soup • 34
Lemon Snaps • 90
Lemon Steam Cake with Blood Orange Sauce • 102
Lemon-Thyme Tea • 112
Pickled Lemons • 72
Pink Lemonade • 108

LIMES

Fried Scallops with Lime and Garlic • 50
Limeade Fizz • 110
Lime and Tomato Relish • 78
Limed Spareribs • 62
Lime Mousse Pie • 96
Lime Soup • 32

ORANGES

Chocolate-Orange Marmalade Brownies • 92
Curried Orange Soup • 36
Frozen Blood Orange Soufflé • 98
Iced Orange Coffee • 108
Mandarin Orange Upside-Down Cake • 104
Navel Orange Waffles with Blueberry Sauce • 22
Orange-Banana Muffins • 18
Orange Butter • 80
Orange-Coconut Custard • 94
Orange French Toast • 24
Pork Medallions with Mandarin Oranges and Cranberries • 64
Veal Chops with Blood Oranges • 66
Vermont Orange Ambrosia • 84

TANGELOS

Tangelo and Sweet Onion Relish • 78

TANGERINES

Beef with Tangerines • 68
Chicken Salad with Tangerines and Red Onions • 42
Tangerine-Cranberry Chutney • 80
Tangerine-Pecan Scones • 20
Tangerine Sorbet • 88

UGLI FRUIT

Baked Fish with Ugli Fruit • 54
Ugli Fruit Caribbean Fish Salad • 40

SOURCES

Citrus is most often available as fresh produce, rather than as packaged, prepared food. You'll find fresh citrus at almost every food store. Organic citrus is often carried by health and natural food stores and, with increasing frequency, at your local supermarkets.

The following are suppliers of quality fresh citrus, organic citrus, and other citrus products. All are available by phone or mail order, unless otherwise noted. Catalog availability, price lists, shipping charges, and product selection can be determined by contacting these suppliers.

American Spoon Foods, Inc.
P.O. Box 566
Petoskey, MI 49770
(616) 347-9030
(800) 222-5886
Citrus preserves and marmalades

Balducci's
424 Avenue of the Americas
New York, NY 10011
(212) 673-2600
(800) 822-1444 outside NY State;
(800) 247-2450 in NY State;
Fax: (212) 985-5065
Fresh citrus, preserves, marmalades

Corti Brothers
5770 Freeport Boulevard, Suite 66
Sacramento, CA 95822
(916) 391-0300
Citrus Marmalades and mincemeats

Crabtree & Evelyn
P.O. Box 167
Woodstock, CT 06281
(800) 253-1519;
Fax: (203) 928-5685
Citrus preserves and marmalades, lemon curd

Cushman Fruit Company
3325 Forest Hill Boulevard
West Palm Beach, FL 33406
(407) 965-3535
(800) 776-7575
Boxed citrus, 'Honeybell' hybrids

Dean and Deluca
560 Broadway
New York, NY 10012
(212) 431-8369
(800) 221-7714;
Fax: (212) 334-6183
Fresh citrus, preserves, marmalades

Esper Products DeLuxe, Inc.
2793 North Orange Blossom Trail
Kissimmee, FL 32741
(407) 847-3726
Citrus jellies, preserves and conserves, tangerine butter

Earl Henrick
279½ Miller Avenue
Mill Valley, CA 94941
(415) 388-5924
Wholesale organic citrus

King Arthur Flour
R.R. #2, Box 56
Norwich, VT 05055
(800) 827-6836;
Fax: (800) 649-3323
Orange oil, citrus extracts, orange flower water

The Larder of Lady Bustle, Ltd.
P.O. Box 53393
Atlanta, GA 30355
(404) 365-9679
Lemon and tangerine sauces

Matthew's 1812 House
Box 15, 250 Kent Road
Cornwall Bridge, CT 06754-0015
(800) 666-1812
Lemon rum cake

Native Farms
332 East 11th Street
New York, NY 10003
(212) 614-0727;
Fax: (212) 614-0719
Fresh organic citrus (local retail only, group sales and wholesale available)

Pete's Spice
174 First Avenue
New York, NY 10009
(212) 254-8773
Citrus oils, extracts, orange water, candied and dried peels

St. Clair Ice Cream Company
One Hanford Place
South Norwalk, CT 06854
(203) 853-4774
Citrus-shaped ice creams and sorbets

Takoma Kitchens
11210 Triangle Lane
Wheaton, MD 20902
(301) 933-3220
Glazed lemon teacakes, strawberry-lime preserves (Mile-high Lemon Pie for local sale only)

Williams-Sonoma
P.O. Box 7456
San Francisco, CA 94120-7456
(800) 541-2233;
Fax: (415) 421-5153
Lemon curd, marmalades, prepared lemons and Seville oranges, citrus juicers and zesters.